GEOLOGY AND PHYSICS OF THE MOON

A Study of Some Fundamental Problems

edited by **Gilbert Fielder,**
Lunar and Planetary Unit,
Department of Environmental Sciences,
University of Lancaster,
Bailrigg, Lancaster, Great Britain.

National space programmes have generated a tremendous public interest in the exploration of the Moon. The fact that man has now been to the Moon, walked across the lunar surface, and collected actual rock samples there, makes discussions of the surface of our satellite all the more realistic and interesting. Since we are in the midst of the American and Russian programmes of lunar exploration, the subject is a living one, with great discoveries being made almost as a matter of course. Whilst no pretence can be made that coverage of recent research material can be complete in a book of this length, some of the most interesting recent advances in our knowledge of the Moon are described. The greater part of the material presented can either not be found elsewhere, or will not be readily accessible to most readers.

The book provides answers to recent questions of great interest, among which are those concerning water on the Moon, the flow of lunar volcanic lavas and the origin of lunar craters. Our knowledge of the detailed structure of the lunar surface layer has advanced rapidly in the last few years, reaching a climax with the return of lunar materials to Earth for analysis. As a result it has been possible to incorporate the findings of authors who have worked with lunar rock specimens loaned by the National Aeronautics and Space Administration of the U.S.A. as well as many of the results of other investigators of lunar material.

Although intended primarily for lunar specialists, the book should be comprehensible to the layman with some interest in astronomy, as the chapters are clearly written and the mathematical content is kept to a minimum.

Gilbert Fielder obtained a B.Sc. (with honours), in Physics and Mathematics, from London University in 1954 and, after further formal training in astronomy and geology, a Ph.D. in lunar studies from Manchester University in 1957. He subsequently obtained a Teacher's Certificate (with full qualifications) at Hull and was awarded an Imperial Chemical Industries Research Fellowship at the University of London.

The author, who has been a Visiting Associate Professor at the University of Arizona's Lunar and Planetary Laboratory in Tucson, a Consultant to the Boeing Aircraft Company at their Scientific Research Laboratories at Seattle and a Lecturer in Astronomy at London University, is at present Reader in Environmental Sciences at Lancaster University and Co-Investigator in the Lunar Apollo Programme of the National Aeronautics and Space Administration.

Dr. Fielder, who is the author or co-author of four previous books, is a Fellow of the Royal Astronomical Society and the Geological Society of London and a member of several international lunar, planetary, and geological scientific societies. In 1954 he was the recipient of the Imperial Chemical Industries Science Medal.

GEOLOGY AND PHYSICS OF THE MOON

A STUDY OF SOME FUNDAMENTAL PROBLEMS

GEOLOGY AND PHYSICS
OF THE MOON

A study of some fundamental problems

Edited by

G. FIELDER

*Department of Environmental Sciences,
University of Lancaster, Bailrigg, Lancaster (Great Britain)*

ELSEVIER PUBLISHING COMPANY Amsterdam – London – New York 1971

ELSEVIER PUBLISHING COMPANY
335 JAN VAN GALENSTRAAT
P.O. BOX 211, AMSTERDAM, THE NETHERLANDS

AMERICAN ELSEVIER PUBLISHING COMPANY, INC.
52 VANDERBILT AVENUE
NEW YORK, NEW YORK 10017

LIBRARY OF CONGRESS CARD NUMBER: 70–151736

ISBN: 0–444–40924–6

WITH 154 ILLUSTRATIONS AND 20 TABLES

PRINTED IN THE NETHERLANDS

Preface and Acknowledgements

The University of London Observatory was a centre for lunar studies during the decade commencing in 1960. In October 1970 the lunar team moved to the University of Lancaster and re-established itself, to take in planetary work, in the Department of Environmental Sciences of that university. It was fitting to round off the team work in London with this research-level publication.

Most of the work reported here was sponsored by the Natural Environment Research Council of Great Britain. We thank the Council for its continuing support in the field of lunar and planetary studies. In addition, much of the work was based on photographs generously provided by the U.S. National Aeronautics and Space Administration.

Chapter 2 is a revised version of a paper with the title "Lava Flows in Mare Imbrium" which was published in August, 1968, by the Boeing Scientific Research Laboratories as their Document Number D1-82-0749. The author of Chapter 3 wishes to acknowledge helpful advice given by P. Ackers and F. G. Charlton of the Hydraulics Research Station, Wallingford. The work described in Chapter 5 was supported under NASA Grant 5601-920-271. The computations described in Chapter 8 were performed using the Atlas Computer of the University of London Institute of Computer Science, the IBM 360/65 of University College London, and the IBM 1130 of the University of London Observatory. One of the authors (L. Wilson) of Chapter 10 thanks Dr. K. Chapman and Dr. P. Whitaker for advice relating to the construction of the proton accelerator, and R. Robson for technical assistance. One of the authors (E. L. G. Bowell) of Chapter 11 acknowledges assistance from Turners Asbestos Cement Company for supplying some of the standard materials used in the thermal conductivity work. Our especial thanks go to Dolores J. Galera for typing from untidy manuscripts and draft copies.

Following the usage suggested by J. F. Vedder, we adopt the adjective "meteoritic" to describe extra-terrestrial material that has fallen to the surface of the Earth, and the adjective "meteoric" to derive from the word "meteoroids"—objects in space. For example, a meteoric crater on the Moon is a crater produced by the impact of a meteoroid.

The reader is asked to note that the plates in this book are presented with north upwards and east to the right. Approximate scales are given on illustrations except in those instances in which the scale changes appreciably from one part of an illustration to another.

G. FIELDER (Editor)
August, 1970

Contents

1

Recent exploration of the Moon

G. FIELDER

Evidence for volcanism from the Ranger programme

Photographic documentation of the Moon has recently leaped forward. Many of the greatest Earth-based photographs of the Moon survived, unchallenged, for over half a century. During that time there was little decrease in the best limit of resolution—typically about 1 km of lunar surface. This backlog of telescopic photographs has now been supplemented by pictures taken from spacecraft sent to the Moon (Fig.1) by the U.S.S.R. (PETROVICH, 1969; Fig.8) and the U.S.A. In 1969 U.S.A. astronauts took the first in situ hand-held camera photographs (Fig.2) of the mare landscape and of individual rocks lying on the lunar surface.

In August, 1964, Ranger 7 transmitted details of the lunar terrain in a part of Mare Nubium renamed Mare Cognitum. For the first time the Moon was seen to be cratered down to sub-metre sizes. There were many more craters than predicted from theories based on extrapolation of the frequency plots of crater numbers as a function of size.

One hypothesis that was offered to account for the large number of small craters was that the excess craters were of secondary impact origin—that is, craters produced by objects ejected from primary impact or volcanic craters at less than the velocity of escape (2.4 km/sec) from the Moon's surface. Advocates of another hypothesis attributed the excess to endogenic craters, that is, to craters of internal origin. Neither hypothesis is beyond doubt and it is possible that both take their share in the explanation of the high number of small craters in Mare Cognitum.

The fact that the area recorded by the Ranger 7 cameras was traversed by faint rays from Tycho or Copernicus lent support to the secondary impact hypothesis: some rays were already known to be zones of increased cratering generally attributed to ejecta

from late impact craters. On the other hand, the integrated mass of objects required to form secondary craters following ejection of objects from a primary crater cannot be greater than the mass excavated from the primary cratering site. Applying a calculation of this sort, KOPAL (1965) deduced that the total volume of ejecta from Tycho was greater than the volume excavated by a factor of at least ten.

Lunar craters commonly cluster to an extent over and above the degree of clustering expected on a random impact hypothesis. Since secondary particles may be ejected in a closely defined spray, the secondary impact mechanism may be invoked to account for non-random clumping of crater centres. Both eumorphic (sharply sculptured, well defined, generally deep) and submorphic (smooth rimmed, poorly defined, generally shallow) craters are in evidence in a cluster of small craters (Fig.3) photographed by Ranger 7. Possibly the secondary bodies were of two categories—coherent blocks and weakly bonded particles, respectively. Both types of material could follow similar trajectories in vacuum: the former would generate eumorphic craters and the latter type of material would produce the submorphic craters.

Another property of the craters in the compact group of craters shown in Fig.3 is the tendency for some of the craters (for example, those lettered *e*) to be aligned. There are three directions of alignment (Fig.4) of these crater chains. The direction of chaining that is most nearly meridional is close to the direction of Copernicus. None of the directions points towards Tycho. Examination of the environs of several large ray craters on the Moon, as well as of model craters produced in the laboratory, leads to the conclusion that secondary craters do tend to line up nearly radially to their parent crater, though there are always exceptions to this rule. Thus, it is possible that at least some of the

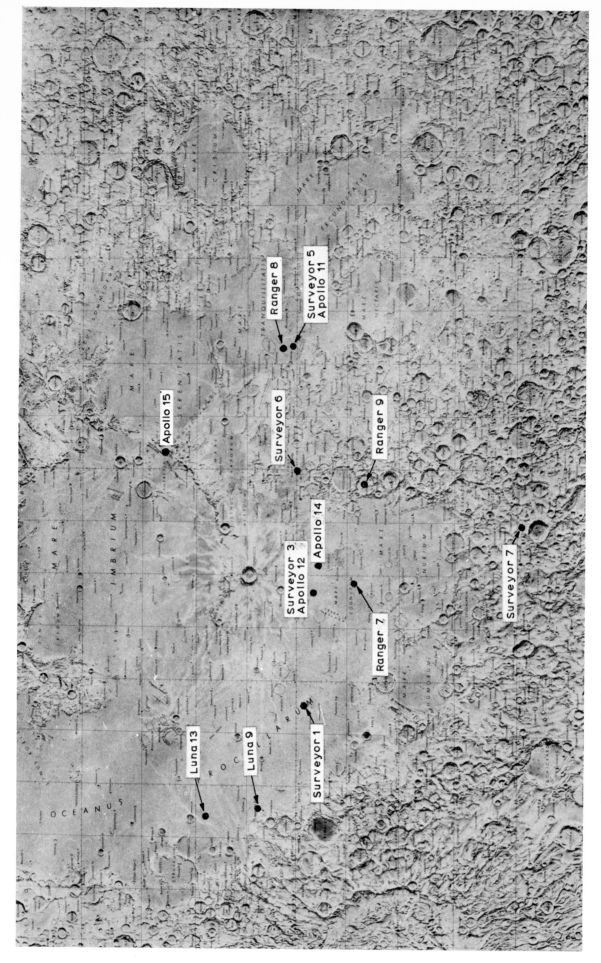

Fig.1. Lunar Earth-side chart, prepared by the National Aeronautics and Space Administration, showing the sites of spacecraft that have significantly increased our knowledge of the Moon.

Fig.2. Lunar landscape at Apollo 11 site (see Fig.1). (NASA photo.)

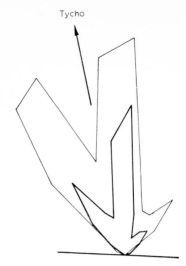

Fig.4. Polar diagram of 136 crater chains and elongated craters (thick polygon), which are members of the cluster of craters seen in Fig.3, and of ridge and depression lineaments (thin polygon) in an area seven times as large as that of Fig.3, but surrounding it. The arrow shows the direction of Tycho.

craters in the north-trending chains are secondary impact craters from Copernicus.

However, the observed chaining directions are paralleled by the respective trends of other lineaments in an extensive field around the cluster of craters (Fig.4). This correlation would be expected if the chain craters were of internal genesis and originated in tectonic fractures. It seems, therefore, that a certain proportion of the craters in the Ranger 7 cluster in Mare Cognitum is endogenic.

This hint, through the Ranger 7 photographs, of lunar craters of internal origin foreshadowed important discussions of lunar volcanism made in the light of

results from later Ranger vehicles and the more advanced spacecraft that followed them.

Before destroying itself in Mare Tranquillitatis, Ranger 8 transmitted photographs under slightly better lighting conditions (lower solar altitude) than those which had prevailed for Ranger 7. Whereas a few dimple-like craters had been seen in the Ranger 7 frames, Ranger 8 photographed several dimple craters one of which is shown in Fig.5. The dimple craters may have been formed by drainage of the regolith— the Moon's ubiquitous but variable veneer of rock fragments—into fractures in the underlying rock. Whereas one can conceive of saucer-shaped submorphic

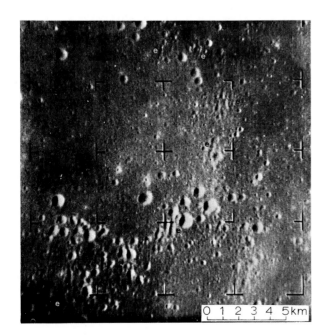

Fig.3. A group of craters photographed by Ranger 7. (NASA.)

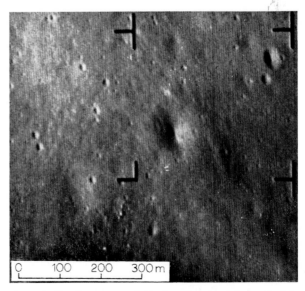

Fig.5. A dimple crater and blocks of rock photographed by Ranger 8. (NASA.)

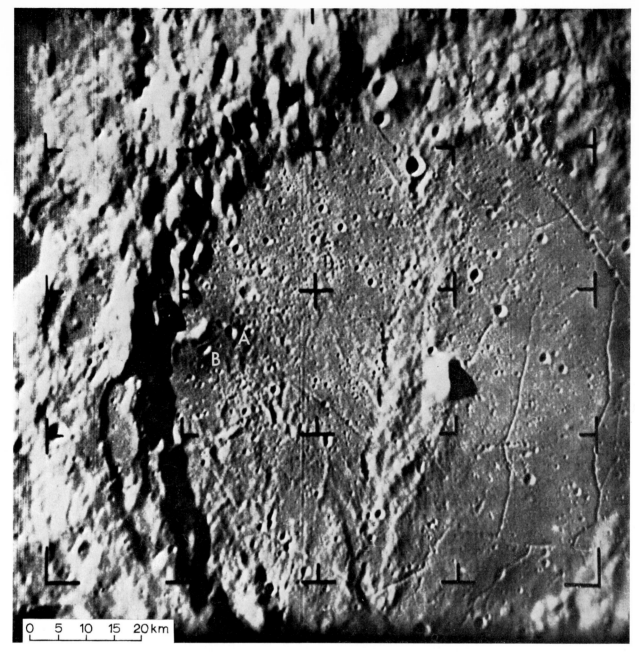

0   5   10   15   20 km

Fig.6. Ranger 9 photograph of Alphonsus. (NASA.)

craters as being denuded eumorphic craters, one cannot generate the convex inner-wall profile of a dimple crater by any such denudation process. KUIPER (1965) found analogues of the lunar dimple craters in terrestrial basaltic lava flows and, hence, was led to explain many of the lunar craters in terms of a drainage or collapse mechanism.

Ranger 8 confirmed that the Moon sported a higher number of craters than had been expected in areas relatively free of major rays from large craters. Numerous blocks of rock and chains of sub-telescopic craters were also photographed by Ranger 8. The

endogenic origin of some of the smaller lunar craters was confirmed but the majority of the craters photographed were still regarded, by most scientists, to be impact craters.

Incontrovertible evidence for craters of internal origin was first seen in photographs of the Alphonsus area (Fig.6) taken in March, 1965, by means of Ranger 9. The detail visible in Fig.6 may be compared with that in the Earth-based photograph of Alphonsus (Fig.7). Two of the well-known dark halo craters, *A* and *B*, that may be seen in Fig.7 in peripheral regions of the floor of Alphonsus, resolve, in Fig.6, into non-

Fig.7. A good Earth-based telescopic photograph of Alphonsus printed on high contrast paper to show the dark halo craters. (Catalina Observatory photo, Consolidated Lunar Atlas.)

circular, rimless craters. They are tied to the rilles—probable collapse features that assume the form of shallow trenches—and the dark haloes associated with the craters coincide with areas where the sharpness of topographic features is appreciably less than in the surrounding areas. McCAULEY (1968) regarded the elongated craters to be eruptive vents that scattered dark ash around them, subduing the relief immediately surrounding each eruptive centre: this interpretation is now generally accepted.

The nature of the regolith as deduced from the Surveyor programme

The Ranger programme improved our knowledge of some lunar details by a factor of more than a thousand. Soft landers such as the Russian Luna 9 (landed February, 1966), and the American Surveyor 1 (landed June, 1966) augmented this factor by another thousand and showed us details of millimetre size (Fig.8 and Fig.9, respectively).

rock is broadly rounded, suggestive of a process of erosion. Indications such as these have now been followed up in detail through the Apollo programme.

Surveyor 3 landed on a 12°–15° slope (Fig.11) on the inner wall of a 200-m sized crater in Oceanus Procellarum. A 12 × 5-cm scoop, attached to an automatic arm, was pushed into the lunar soil for the first time. The subsequent trenching operation was observed via the on-board television camera and the upper parts of the regolith were seen to behave like damp sand. The trench excavated by the surface sampler of Surveyor 7 is depicted in Fig.12. Evidently the particles forming the regolith are capable of sustaining 90° slopes.

Surveyor 5 landed just within the crest of a 9 × 12-m crater in Mare Tranquillitatis, Surveyor 6 near to a mare ridge in Sinus Medii, and Surveyor 7 set down successfully in a relatively blocky highland unit (Fig.13) —probably a flow unit—some 20 km from the northern rim of Tycho.

These last three Surveyor craft carried ^{242}Cm sources used to irradiate selected parts of the lunar

Fig.8. Lunar landscape photographed by Luna 9 (see Fig.1).

All the Surveyor craft carried television cameras. Surveyor 1, primarily an engineering mission, landed in the ring structure Flamsteed P, discussed in Chapter 7. The exceptionally sharp panoramic and close-up photographs transmitted from Surveyor 1 were of great scientific interest: one whole rock (Fig.10) is fractured, pitted, and clearly shows brightness variations indicative of compositional differences. The surface of the

surface with α-particles. The spectra of α-particles and protons scattered from the lunar rocks provided the first indications of the kind and quantity of elements present at the surface of each test area. Analysing these spectra, TURKEVICH et al. (1967, 1968, 1969) and PATTERSON et al. (1970) found only slight differences between the atomic percentages of the rocks at the Surveyor 5 and 6 sites, but larger differences between

Fig.9. Pitted block and small particles photographed by Surveyor 1 (see Fig.1). Note accumulation of fine material near base of block. (NASA/JPL computer enhanced photo.)

Fig.10. Fractured block, about 50 cm long, showing textural and tonal differences. (Surveyor 1 photo, NASA.)

Fig.11. Astronaut from Apollo 12 lunar module (on skyline) examining the Surveyor 3 craft which made a soft landing two and a half years earlier. (NASA photo.)

Fig.12. Trench scooped by automatic equipment carried on the Surveyor 7 craft. (NASA/JPL computer enhanced photo.)

Fig.13. Blocky terrain, photographed from Surveyor 7, of the rim unit of Tycho. Note blocks differ, between themselves, in roundness. (NASA.)

these two analyses, on the one hand, and the Surveyor 7 analyses, on the other. In particular, at the Surveyor 7 site, iron is about half as plentiful and aluminium about twice as plentiful as at the mare sites. Titanium—high in the mare sites—is practically absent at the Surveyor 7 highland site.

Surveyors 5, 6 and 7 also carried magnets which were used to show that the magnetic susceptibility of the particles of the regolith was similar to that of terrestrial basalts.

The thickness of the regolith at each Surveyor site was estimated by observing, on the television pictures, the depth of the smallest neighbouring crater from which apparently coherent blocks of rock had been ejected. The theory is that a shallow enough crater would have been scooped from the regolith alone, whereas a crater deeper than the thickness of the regolith would have been excavated from the relatively rigid underlying rock. In this way it has been deduced that the regolith is 1 or 2 m thick at the Surveyor 1 site, 1 or 2 m at the rim of the Surveyor 3 crater, but more than 10 m at the centre of the floor of the crater, less than 5 m at the Surveyor 5 site, and 10–20 m at the Surveyor 6 site. Estimates at the Surveyor 7 site are ambiguous because the many blocks of rock on the rim unit (Fig.13) of Tycho make this method of determining the thickness of the regolith unreliable.

Estimates of the crushing strengths of rock fragments were made by loading specimens picked up in the jaws of the surface samplers carried by Surveyors 3 and 7. One highland rock weighed and measured at the Surveyor 7 site seemed to have a density of 2.8 $\pm$ 0.3 g/cm^3. Accurate determinations of the density of crystalline rocks returned from the Apollo 11 mare site gave mostly 3.1–3.5 g/cm^3. Possibly the lunar highland rocks have a generally lower density than the mean density of the rocks forming the maria. This thesis has been developed by PATTERSON et al. (1970).

Lunar Orbiter photographs and the discovery of mascons

The Surveyor programme led to confirmation of earlier observations that it would be safe for man to set down a module on the Moon and step out across the surface. High-resolution maps of the specific landing sites proposed by the U.S.A., and photographic coverage of both hemispheres of the Moon, were provided through the highly successful Lunar Orbiter series of camera-equipped vehicles. Broadly speaking the aim of Orbiters 1, 2 and 3 was to scrutinize, photographically, possible manned landing sites in preparation for the Apollo programme. These sites were located principally in the equatorial region of the Moon's near side to facilitate landing operations and communications with Earth. The early Lunar Orbiters were placed in low-altitude lunar-equatorial type orbits. Orbiter 4 was placed in a high-altitude orbit inclined at 85° to the Moon's equator to obtain as complete a photographic survey of the Moon as possible. Finally, Lunar Orbiter

Fig.14. Lunar Orbiter 5 photograph of parts of Alphonsus. (NASA.)

5, also in a low-altitude, high-inclination, orbit, was made to photograph some near-side areas of outstanding scientific interest (the area depicted in Fig.14, for example) as well as being programmed to record areas previously not photographed, or not well photographed, on the far side.

All the Lunar Orbiters took both medium-resolution (Fig.14) and corresponding high-resolution (Fig.15) photographs. The dark halo craters *A* and *B*, mentioned earlier, may be studied in detail in these photographs. The best ground-based resolution was about 1 m, attained by Orbiters 2 and 3. Many Lunar Orbiter photographs will be found in this book; their scientific importance in lunar studies is enormous, the discoveries to which they have already led innumerable, and the wealth of detail they portray presently unaccountable. It will be many years before all these data

are digested—and then by only a few scientists.

One important side-product of the Lunar Orbiter programme was the discovery of "mascons". Using the radio transmissions and Doppler data from Orbiter 5 when it was in orbit around the Moon, MULLER and SJOGREN (1968) detected anomalies in the calculated radial accelerations and conjectured that the residual positive accelerations were produced by mass concentrations below the surface of certain of the maria. One of the largest of these gravity anomalies occurs over Mare Imbrium—a classical example of a circular type of lunar mare—and will be discussed further in Chapter 2. The question as to whether mascons are dense objects buried well beneath marial surfaces or whether mascons are simply accumulations of relatively dense lavas in sheet or lenticular form close to the surface has not been resolved.

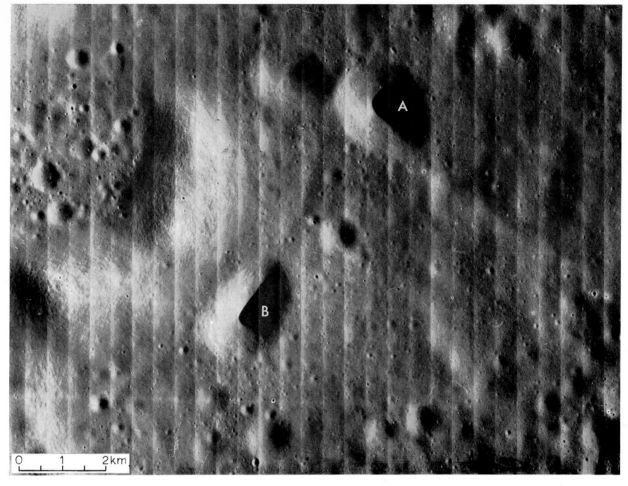

Fig.15. Close-up photograph of three dark halo craters shown in Fig.14. (NASA.)

Apollo analyses and the origin of the Moon

This logistic programme of unmanned space vehicles set the stage for the U.S.A. manned lunar landings. Astronauts first took high quality photographs (for example, Fig.16, 17) of interesting areas. Early stereo photographs (Fig.18) are of assistance in interpreting details. Some of the scientific results from the first two landing missions—Apollos 11 and 12— are startling.

Because chondrites had appeared to be the most common type of meteorite it had been widely supposed (for example, by UREY, 1968) that the bulk of the Moon was chondritic. The Apollo results proved that the Moon as a whole was not of chondritic composition. Crystalline rock samples returned to Earth for mineralogical and petrological analyses revealed that the rocks of the maria were volcanic and that they had been derived by differentiation from clinopyroxene-type magmas at depth (for example, see O'HARA et al.,

1970). Strangely enough, the composition of the Moon is unlike the composition of any meteorites or that of the Earth's crust or mantle; and this seems to dispose of the theory that the Moon was born from the Earth by fission.

The lunar rocks contain more radioactive elements than chondrites. UREY (1960, 1969), in particular, had argued against the hot Moon theory; yet, taken together with the Apollo radioactive and age data, thermal calculations (ANDERSON and PHINNEY, 1967; McCONNELL et al., 1967; FRICKER et al., 1967) now indicate that the Moon must have melted at depth at some time in the past.

Another surprise to many is that the lunar breccias, and the regolith particles smaller than 1 cm in diameter—the so-called fines, from which the breccias may have been formed by impact-generated shock—contain only of the order of 1 % meteoric material. Evidently, the Moon is not accumulating meteoroids to a significant extent. The fines contain a few percent

Fig.16. Part of floor and central mountain complex of a lunar far-side crater photographed by an astronaut from Apollo 10. Arrows mark flows from mountains. (NASA.)

Fig.17. Lunar far-side crater chains and other lineaments photographed by an astronaut from Apollo 8. (NASA.)

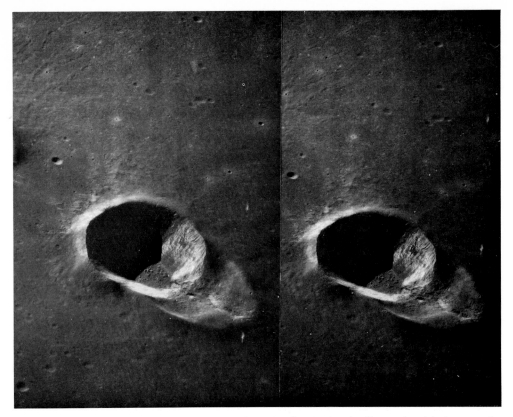

Fig.18. Messier A photographed by an astronaut from Apollo 11. Fine details may be picked out on the craters' walls and floors. (Pair of stereo photos, NASA.)

of anorthosite which cannot have been derived locally; and the fines from the Apollo 11 site differ chemically from those collected at the Apollo 12 site; so thorough mixing of fine particles on a Moon-wide basis has not occurred in the exposure times of 10^6 years to a few 10^8 years deduced from studies of cosmic ray nuclides in whole rocks. It has been widely conjectured that the anorthosite (density only 2.9 g/cm^3) is material ejected by impact or volcanic processes in the lunar highlands. It is similar in its overall composition to that of highland terrain analysed using Surveyor 7; and the density lies in the range of uncertainty of that of a rock measured using Surveyor 7. Mixing of regolith material is proceeding, but only very slowly.

COMPSTON et al. (1970) contend that the Apollo 11 crystalline rocks are of two distinct chemical types. It is possible that they separated out as a result of at least two periods of melting. In principle, lunar rocks could be melted through early accumulation processes, dissipation of tidal energy in the Moon, radioactivity or the kinetic energy of major impacts. The termination of the last melting phase of the rocks—whichever of these processes produced it—must have set the measured "ages" of $3.7 \cdot 10^9$ years to the Apollo 11 crystalline rocks from Mare Tranquillitatis and $3.4 \cdot 10^9$ years to the Apollo 12 crystalline samples from Oceanus Procellarum.

In addition to rock studies, the Apollo programme involved physical experiments that proved to be of great value. For example, the laser retro-reflector experiments are supplementing our knowledge about the figure and orbit of the Moon—data vital to assist us in understanding the origin of the Moon. Other instruments now on the Moon are being used to inform scientists about the physical conditions of the lunar environment.

Perhaps the most unexpected results from landed instruments stemmed from the seismometers. The Apollo 12 seismometer recorded two man-made impacts: those of the module "Intrepid" and of the Saturn booster rocket which had been used to launch Apollo 13. The records were similar, but were most unusual because they took the form (Fig.19) of a slow rise to maximum amplitude followed by a surprisingly drawn out decay lasting at least ten times as long as

Fig.19. One key type of lunar seismogram. (From LATHAM et al., 1970.)

expected from experience of most terrestrial records. Numerous miniaturised records of the same general form have been detected by both the Apollo 11 and 12 seismometers. LATHAM et al. (1970) think that these records represent meteoric impacts. The reverberations seem to indicate that the upper parts of the lunar maria are effective scatterers of seismic energy which, nevertheless, is not damped out rapidly. A Moon, riddled with interfaces—such as those that might be provided by a plethora of lava tubes or welded fractures—might account for these observations.

With the great wealth of multi-disciplinary data now emerging from the Apollo programme we are merely on the doorstep to the history of the Moon and of the Earth–Moon system. Other samples of geologically different parts of the Moon are required if we are to understand the Moon better. Inaccurate as it undoubtedly is, our present picture is that of a Moon with a small iron or iron sulphide core within a thick, gabbroic "mantle". Close to the surface there is an anorthositic "crust", in places punctured by basaltic extrusives stemming from partial melting processes in the "mantle".

Developing an earlier theory of E. J. Öpik, RINGWOOD and ESSENE (1970) consider the Moon itself to have formed from a ring of planetesimals around the Earth. Silicates falling from this ring evaporated in the Earth's primitive atmosphere. As the atmosphere dissipated, the silicates precipitated and accumulated to form the Moon. In this way the Moon was built of material having a lower mean density than that of the Earth.

References

ANDERSON, D. L. and PHINNEY, R. A., 1967. In: S. K. RUNCORN (Editor), *Mantles of the Earth and Terrestrial Planets*. Wiley, London, p.113.

COMPSTON, W., ARRIENS, P. A., VERNON, M. J. and CHAPPELL, B. W., 1970. *Science*, 167: 474.

FIELDER, G., 1968. In: S. K. RUNCORN (Editor), *Mantles of the Earth and Terrestrial Planets*. Wiley, London, p.461.

FRICKER, P. E., REYNOLDS, R. T. and SUMMERS, A. L., 1967. *J. Geophys. Res.*, 72: 2649.

KOPAL, Z., 1965. *Boeing Report*, D1-82-0475.

KUIPER, G. P., 1965. *Sky Telescope*, 29: 293.

LATHAM, G. V., EWING, M., PRESS, F., SUTTON, G., DORMAN, J., NAKAMURA, Y., TOKSÖZ, N., WIGGINS, R., DERR, J. and DUENNEBIER, F., 1970. *Science*, 167: 455.

McCAULEY, J. F., 1968. *Am. Inst. Aeronaut. Astronaut.*, 6: 1991.

McCONNELL JR., R. K., McCLAINE, L. A., LEE, D. W., ARONSON, J. R. and ALLEN, R. V., 1967. *Rev. Geophys.*, 5: 121.

MULLER, P. M. and SJOGREN, W. L., 1968. *Science*, 161: 680.

O'HARA, M. J., BIGGAR, G. M. and RICHARDSON, S. W., 1970. *Science*, 167: 605.

PATTERSON, J. H., TURKEVICH, A. L., FRANZGROTE, E. J., ECONOMOU, T. E. and SOWINSKI, K. P., 1970. *Science*, 168: 825.

PETROVICH, G. V. (Editor), 1969. *The Soviet Encyclopaedia of Space Flight*. MIT, Moscow.

RINGWOOD, A. E. and ESSENE, E., 1970. *Science*, 167: 607.

TURKEVICH, A. L., FRANZGROTE, E. J. and PATTERSON, J. H., 1967. *Science*, 158: 635.

TURKEVICH, A. L., FRANZGROTE, E. J. and PATTERSON, J. H., 1968. *Science*, 162: 117.

TURKEVICH, A. L., FRANZGROTE, E. J. and PATTERSON, J. H., 1969. *Science*, 165: 277.

UREY, H. C., 1960. *J. Geophys. Res.*, 65: 358.

UREY, H. C., 1968. In: *International Dictionary of Geophysics*. Pergamon, London, p.3.

UREY, H. C., 1969. *Bull. Atomic Sci.*, 25: 46.

2

Lava flows and the origin of small craters in Mare Imbrium

G. FIELDER AND J. FIELDER

Discovery of flows

In the past few years it has been established that at least some of the lunar maria are composed of lavas. BALDWIN (1949) argued that the lavas flowed for distances of the order of 1,000 km—for instance, from Mare Imbrium to Mare Serenitatis—whereas FIELDER (1965) reasoned that mare ridges were lavas that had been extruded from fissures and then flowed across the surface for distances of the order of 10 km. Most authors regarded the postulated flows to be of basaltic type lavas; but whereas BALDWIN (1949), FIELDER (1965) and others considered the lavas to be derived internally, UREY (1952) and others argued that the lavas were generated by the impact of the low-velocity planetesimals which, they believed, sculptured the respective mare basins and their surrounding rings of mountains.

Important steps forward, reported by KUIPER (1965), were taken by E. A. Whitaker and R. G. Strom. Whitaker secured full-Moon telescopic photographs in the infrared and ultraviolet which revealed colour differences between certain parts of the maria. With Strom, he assembled a large number of near-terminator telescopic photographs in yellow light which showed lobate forms characteristic of lava flows. Furthermore, the boundaries delineating these fronts were found to coincide with the curves drawn to separate regions of different colour.

Strom mapped (FIELDER and FIELDER, 1968; Fig.1) the units thus identified in Mare Imbrium. It is clear from Fig.1 that several of the units are of flow-like form and seem to have their origin in mare ridges.

Because of the great scientific interest in the Mare Imbrium flows, Lunar Orbiter 5 was commanded to photograph the rectangular area, identified in Fig.1, in both medium- (Fig.2) and high-resolution (Fig.3).

Stratigraphic overlap of units may be examined to show that unit f1 was present before unit f2 was laid down and that the unit f3 was the latest to form in the region.

Geologic description of the flow f3

The flow f3 is centred at about 23°W 31°N, in the southwest quadrant of Mare Imbrium, and is about 130 km long and 20–50 km wide. Details down to the 50-m size may be identified readily on the medium-resolution frame number 159 and down to 9 m on the corresponding high-resolution photography, the coverage of which is indicated by means of broken lines in Fig.4. A general geologic study may best be pursued using the medium-resolution photography. Useful supplementary information stems from the high-resolution coverage and from Earth-based photography.

The readily identifiable fronts of the flow f3 are shown in Fig.5 and the most conspicuous of several flow units have been identified and lettered *fu1* through *fu6* in Fig.4. These flow units issue from the bases of some of the steepest fronts and are probably the result of fluid pressure overcoming the confining pressure of the congealing front of the principal flow. Each flow unit has its own front but it is not as high, or as steep, as the principal front. Viewed in plan, the fronts of both the main flow and of the flow units are coarsely lobate—a form that characterises the terminus of a liquid that has flowed across the undulating surface of a gravitating body.

The thickness of f3 can be inferred from the thickness of its fronts which cast measurable shadows. Shadows on a medium-resolution negative were measured in the direction of the solar azimuth and estimates of shadow lengths were made, also, from a high-reso-

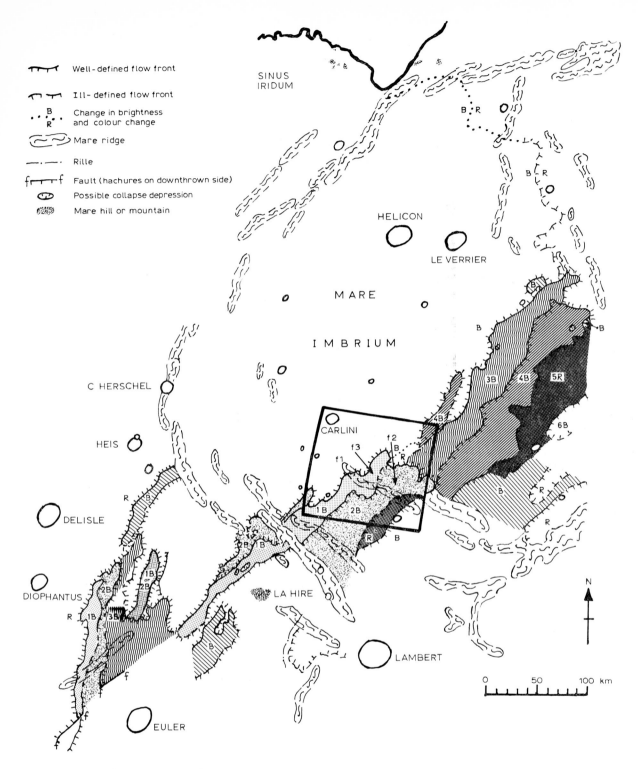

Fig.1. Flows in Mare Imbrium. The enclosure isolates the area shown in Fig.2. (Mapped by R. G. Strom.)

lution print (Fig.3). The angle of incidence of sunlight was taken to be 72°. The steepest parts of the fronts attain altitudes of up to 25 m. However, the high-resolution photographs show that there is commonly a gentle slope starting at the foot of the steepest part of a front. Whatever the origin of this gentle toe—which may be of talus—the indication is that the original height of the flow fronts exceeded 25 m. The frontal thickness of f3 is probably less than the maximum height (~40 m) of the fronts of lava flows mapped

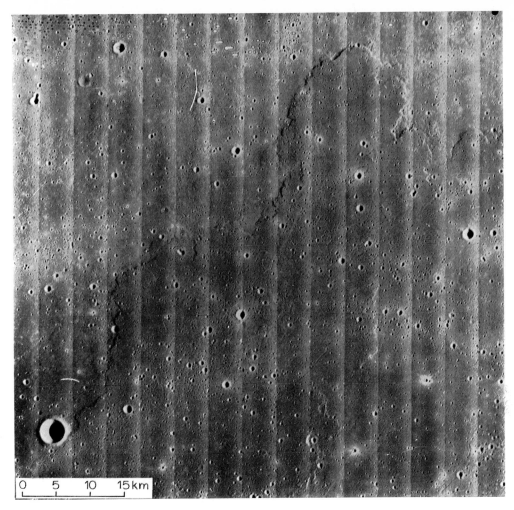

Fig.2. Medium-resolution Lunar Orbiter 5 photograph of the geologic unit f1 and the flows f2 and f3 identified in Fig.1. (NASA.)

by STROM and FIELDER (1968; see also Chapter 5) in the region of Tycho.

Although intensely cratered the surface of the unit f3 is, on a broad scale, level to a remarkable degree. There is no conspicuous flow ridging or arching and fracturing of the surface such as one sees in the lava flows (see Chapter 5) associated with Tycho, for instance. It is shown, later, that the present surface of f3 is not the original surface of the flow.

Numerous shallow channels (Fig.3, 5) traverse f3 and show a tendency to fan out toward the best developed fronts at the distal end of the flow and become orthogonal to them. The channels are up to 180 m wide and the longest continue for about 8 km. The trace of a channel has a radius of curvature which is large compared with the width of the channel: there are no sharp meanders that match those of the highly sinuous rilles (see Chapter 3). Away from the distal end of the flow, most channels tend to parallel one another. In the map (Fig.5) long channels are the more

numerous. Shorter channels and crater chains (which have not been distinguished separately) show a strong tendency, like the lava fronts, to follow the two principal directions[1] of the grid system. Whereas the short, linear channels and crater chains may be fracture-controlled, the long, curving channels tend to break away from and cut across the principal grid directions and, hence, appear to be rootless. One of the best developed channels, *cb* in Fig.5, was covered by the high-resolution photography which shows that the channel may possibly bifurcate at *cb* to form sibling channels. The probable directions of flow are indicated with arrows in Fig.6; thus, *cb* may be described as a possible "distributory".

Distributories are found in lava tubes and lava drainage channels on Earth. Levees characteristic of one class of old lava channel are absent from the

[1] Linear topography trending in these directions is widespread on the Moon and, together with trends in other directions, has been termed the lunar grid system (see, e.g., FIELDER, 1961).

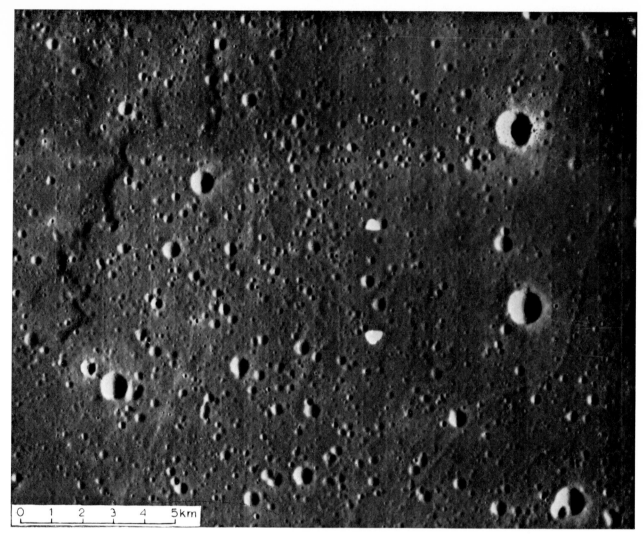

Fig.3. High-resolution Lunar Orbiter 5 photograph showing part of the flow f3 and its front. (NASA.)

mapped channels, however. Rather, the channels are punctuated, not uncommonly, by rimless craters which assume the form of collapse depressions. Several of the channels appear to terminate and then continue beyond a stretch of terrain which shows no expression of a channel.

Precisely these phenomena are found in the case of partially collapsed terrestrial lava tubes (Fig.7; Chapter 3). An hypothesis which is consistent with all the observations is that the channels were formed by the partial or complete collapse of the surface into lunar lava tubes which acted as feeders of lava, transporting it from the main fissure or fissures to the active fronts of the flow.

Crater counts in the region

The craters on the flow f3 deserve special attention since they are in excess (see Fig.2) of those on the other major unit, f1, in the region mapped. Since f3 is stratigraphically the younger of the two units it should carry fewer primary impact craters per unit surface. The writers have assumed that all the craters are primary impact craters and have constructed various regolith models but none of them accounts for all the facts. Either the excess of craters on f2 is due to secondary impacts or it represents a minimum of internal cratering.

A positive print measuring 146 × 119 cm² was prepared from the medium-resolution Orbiter photograph and the scale was determined by measuring distances between the centres of the three smallest craters—Carlini B, H and K—with listed coordinates (ARTHUR et al., 1964). Concentric circles of measured radii were scribed on a sheet of clear acetate and this template was used by one of us (J.F.) to count craters, lying between selected size limits, in the sample areas M1 (in unit f1) and M2 (in the flow unit f3, Fig.4).

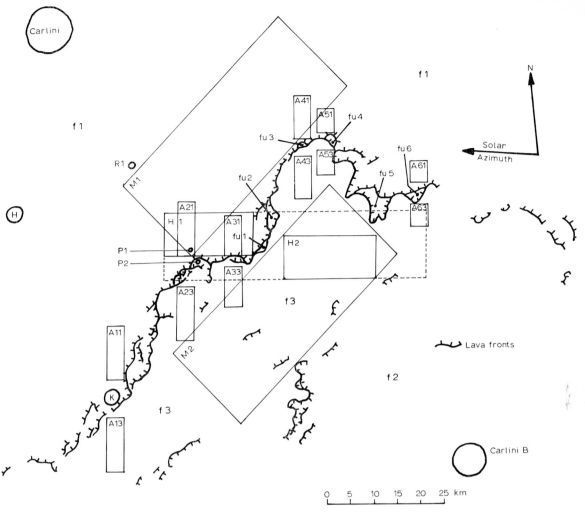

Fig.4. Key map of the area covered in Fig.2 showing the sample areas used for the first crater counts.

It was found that there were 2.2 times as many craters between 333 m and 500 m in diameter and 2.1 times as many between 500 m and 1,000 m in diameter in the flow area *M2* as in the area *M1*, off the flow. The smallest diameter measured in each *M*-area was equivalent to a template diameter of 3.72 mm: this was large enough to ensure that there were no substantial observational losses of craters.

Further counts were made by one of us (G.F.) in six pairs of strips (*A11, A13* through *A61, A63*) on opposite sides of the lava front and in such a way as to avoid: (*a*) flow units; (*b*) zones of possible secondary craters around large craters (unless the large crater was symmetrically placed with reference to the opposing strips); and (*c*) any photographic defects. The long edge of every strip was located at a measured distance within the artificial joins between Orbiter framelets, since the photographic quality was highest away from the joining edges of each framelet. In this way the comparison counts, on and off the flow, were restricted to areas of virtually identical photographic quality—an important consideration when craters almost as small as the effective limit of resolution were to be counted.

Whilst craters larger than 225 m are about twice as numerous in the unit area of f3 as in the unit area of f1, craters smaller than this limit generate approximately equal number densities in the respective units f1 and f3.

Completeness of counts (by J.F.) of the smaller craters off and on the main flow, respectively, was ensured by selecting two approximately equal areas *H1* and *H2* (Fig.4) on the high-resolution coverage of the region. The scale was determined from that of the medium-resolution print, by identifying craters; and a template was constructed to accommodate craters down to 160 m in diameter, or 4.37 mm of high-resolution print. At this size of template observational losses were thought to be small.

Each of the three sets of counts shows, independently, that there is an abrupt increase in the

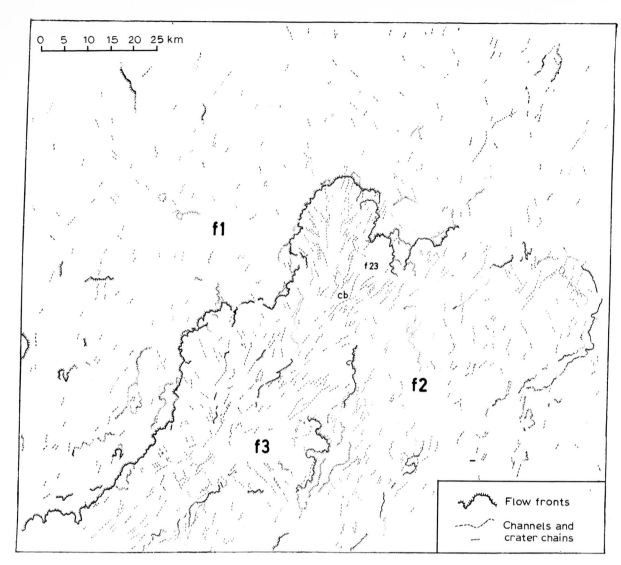

Fig.5. Geologic sketch map of the area covered in Fig.2. (Mapped by J. B. Murray.)

number density of craters larger than about 200 m in diameter when the principal flow front is traversed and the region of the flow f3 is entered. Although some preferential secondary cratering may contribute to this change the effect of secondaries must be small since the discontinuity in number density is essentially bound to the mapped fronts of the flow. There is no reason to impose such a strict and peculiar areal limitation on the distribution of secondary craters in this part of Mare Imbrium. Thus the excess of craters on the flow is due essentially to craters of other than primary or secondary impact origin.

Completely independent counts of craters were made by J. B. Murray (personal communication, 1970). In this survey, the diameters of craters were measured

using a straight edge. Using the same medium-resolution Orbiter print enlarged to 146 × 119 cm² he counted all the craters larger than 267 m in diameter in the bounds (as presented in Fig.6) of f1, f2 and f3, taken separately, but excluded the small unit f23 (Fig.6) which R. G. Strom (Fig.1) had taken to be part of f2 and FIELDER and FIELDER (1968) had taken to be part of f3.

All the data for units f1 and f3 are plotted as a ratio of number densities in these units, against crater size, in Fig.8. In all cases—involving three different observers, two different methods of accumulating data and several different sample areas—there are significantly more craters larger than about 200 m in diameter per unit surface of f3 than of f1.

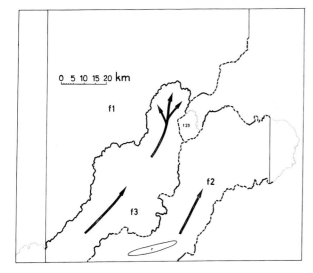

Fig.6. Key map showing the boundaries of the sample areas f1, f2 and f3 used in the counts of J. B. Murray. Craters in the areas f23 and r (a heavily cratered ray) were not counted.

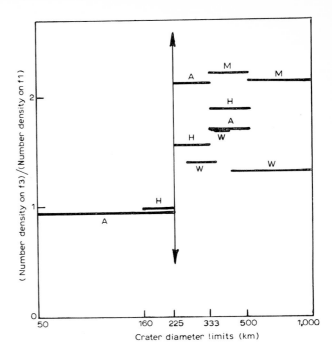

Fig.8. Ratio of number densities of craters of different sizes in different specimen areas of the geologic units f1 and f3. Symbols A, M and H are described in the text and in Fig.4. The symbol W refers to the whole area counts of J. B. Murray.

Fig.7. Tube, formed by molten lava, in Oregon Lava Caves State Park. (Oregon State Highway Commission photo.)

The nature of the flow f3 and the origin of mascons

Evidently the series of flows (Fig.1) observed by R. G. Strom were erupted at different times from the same fissure system. Essentially similar fissure eruptions occur in Iceland and elsewhere. Flows of this type are invariably basaltic, of low viscosity, and cover large areas.

Possibly f1 is an early flow. The larger craters on f3 tend to be sharper than those on f1, with the exception of a group of ray craters, which are among the sharpest in the region. Using circular and elliptical templates, J. B. Murray assessed the shapes of all craters larger than 267 m in diameter in the region identified in Fig.6. He classed crater lips, seen in plan, as circular, elliptical or irregular. The results (Table I, II) show that the craters in the unit f3 are closer, on average, to the circular form than are the craters in units f1 and f2. Tectonic deformations may have devel-

TABLE I

SHAPES OF CRATERS BETWEEN 357 AND 446 M IN DIAMETER

Unit	No. of craters	Percent of craters		
		circular	elliptical	irregular
f1	132	71	22	7
f2	106	64	29	7
f3	102	83	14	3

TABLE II

SHAPES OF CRATERS LARGER THAN 446 M IN DIAMETER

Unit	No. of craters	Percent of craters		
		circular	elliptical	irregular
f1	148	78	20	2
f2	113	78	15	7
f3	89	87	11	2

oped more in the older units, thus leaving the craters in the most recent unit f3 relatively undeformed with respect to an original circular shape.

The material of f3 seems to have behaved like a low-viscosity lava, for, although tending to the lobate form, the flow fronts conform, more specifically, to the directions of the linear undulations in the terrain labelled f1; and the front becomes less high when it abuts against the outer slopes of a crater such as Carlini K (Fig.4). Again, the comparison of a once-viscous looking flow (fig.17 of Chapter 5) near Tycho with the flow f3 leads one to suggest that the Mare Imbrium flow is the more basic in its composition.

Because of the lack of atmosphere, it is reasonable to suppose that the lavas of f3 would outgas rapidly and freeze in the form of a highly porous rock froth (FIELDER et al., 1967) which would act as an excellent thermal insulator of the liquid lava in layers or tubes beneath it. Because of the low lunar gravity the roof of the flow would have a total thickness that would be expected to exceed the roof thickness of a chemically identical terrestrial lava. Thus, the flow thickness of 25 m or more does not preclude a lunar lava of low viscosity. For all these reasons, it is argued that the flow f3 is very probably of a basaltic type of lava.

Together with the other mapped flows in Mare Imbrium, f3 probably discharged some $2 \cdot 10^3$ km^3 of basaltic lava. Judging by their areal coverage, the mapped flows probably represent less than 10% of the total volume of lava that is contained by the present Mare Imbrium. The eruption of $\sim 10^4$ km^3 of lava could have derived from a magma chamber or chambers in Mare Imbrium.

MULLER and SJOGREN (1968) observed positive radial acceleration residuals of Lunar Orbiter 5 as it passed over certain of the maria. This led them to propose that there were concentrations of mass ("mascons") in certain places in the Moon. Anomalies of this sort are associated with the subcircular maria Imbrium, Orientale, Serenitatis, Crisium, Smythii, Nectaris and Humorum; and with Grimaldi, Sinus Aestuum, Mare Humboldtianum, an area north of Theophilus and an area east of Vendelinus. All these positive anomalies are found to be over terrain that is predominantly of the mare type. It is possible to explain the gravity anomalies qualitatively in terms of a pluton, degenerate lopolith, or in terms of sheets of exceptionally dense lavas spread over certain of the maria.

The basalts returned from Mare Tranquillitatis—not a mascon site—have densities of around 3.0 g/cm^3. The whole Moon cannot consist of this kind of rock since, under the conditions prevailing in the Moon at depths of ~ 100 km, it would transform to eclogite, which has a density of 3.7 g/cm^3, whereas the mean density of the Moon is only 3.3 g/cm^3. Possibly the total thickness of lavas in Mare Tranquillitatis is small compared with that in a mascon area.

O'HARA et al. (1970) consider mascons to be generated by alternate layers of high-density iron–titanium oxides and less dense minerals from which the heavier minerals separated following fractional crystallisation in a lava "lake". A 10-km deep lava lake with 10% TiO$_2$ in the initial liquid could provide enough basal concentration of high-density material to account for the mascons. Subsequent lavas would be extruded from depth under hydrostatic pressure, since the density of the liquids at depth would be less than the densities of the solid phases on or near the surface. This model seems to have the potential of meeting our present observations. Detailed gravimetric analyses will be required to determine if an intrusive, or near-surface, concentration of mass is able to account quantitatively for the observations of MULLER and SJOGREN (1968).

The significance of the eumorphic craters in the region

One class of crater in the region mapped is particularly conspicuous. It is sharply sculptured and the rim forms a thin, crescental shadow outside the crater as well as an internal shadow that is longer than the shadow in other craters of the same size. Many of these sharp craters are unusual, too, in that they are "double-walled" (Fig.9) and many of them contain a large block or blocks (Fig.10). The largest of them, at least, are surrounded by a bright ray halo.

The craters are of particular importance in diagnosing the nature of the latest phases in the formation of the lunar surface in the mapped region. Using the medium-resolution negative, thirteen such eumorphic craters upwards of 200 m in diameter were counted to the south and east of the principal flow front and eighteen in an approximately equal area of the unit f1. All thirteen craters behind the front had bright ray haloes and three of them had blocks placed centrally in them. In unit f1, sixteen of the eighteen craters were centres of bright haloes and five craters contained central blocks.

Counts on both the medium- and the high-resolution coverage show that the eumorphic craters with upturned rims are roughly equally numerous in equal areas to either side of the principal flow front of f3. They appear to be randomly distributed and, otherwise, bear the characteristics of impact craters; possibly those

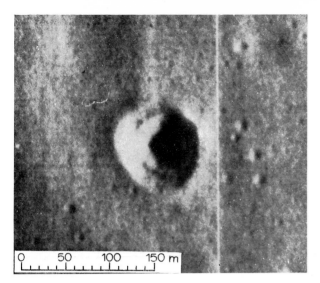

Fig.9. Double crater with deep inner crater. (NASA photo.)

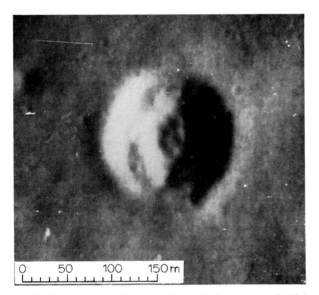

Fig.10. Double crater with central blocks in inner crater. (NASA photo.)

containing blocks are secondary impact craters.

It is of great interest to note that, with respect to all the eumorphic craters larger than 55 m in diameter in the region, 70% of the craters on the flows f2 and f3 are "double" and 69% of the craters off the flow, in unit f1, are likewise double. A further striking observation is that, whereas only 8% of the craters on the flows f2 and f3 contain central mounds, or blocks, by contrast as many as 58% of the craters off the flow, in unit f1, contain such central blocks.

As in the case of the large, eumorphic craters the numbers of these smaller craters per unit area are essentially equal on and off the flow. The smaller eumorphic craters also appear to be randomly distributed in both geologic units. The illumination is not

high enough to reveal whether all these small craters are bright ray centers but their rims have characteristic lips and are similar to those of the larger eumorphic craters which are seen to be generally bright. For these reasons the craters are probably recent impact craters.

Of the double craters, most are concentric. A few have their inner craters eccentrically placed and may have been formed by impacts at large angles of incidence. The lowest parts of the floors of most of the other, more commonplace, craters in the region are not covered by shadow but, in general, the inner crater of the double objects (Fig.9; D in Fig.11) in the flow f3 are so deep that shadow covers up to two thirds of their diameter. This is not so for the majority of the double craters off the flow; here the inner craters (when not replaced by blocks, as in Fig.10) are shallow.

Microscopic examination of the double craters and measurements of the shadows in them shows that the craters have a bench at an estimated average depth of about 6 m. This depth seems to be approximately the same for craters of different size and for those located in both the units f1 and f3. The benches give the craters their double appearance, which is strikingly like that of the experimental craters formed by OBER-BECK and QUAIDE (1967, 1968) in two-layered regimes in which the lower layer is better bonded than the upper layer.

It is generally recognised that in situ impact brecciation and churning, or large craters ejecting rock deposits such as ash, are the agents responsible for the surface modification. Because, in the region mapped, there are colour differences between the different geologic units, and because the thickness of the surface layer seems to be nearly constant, the former mechanism is to be preferred in accounting for the regolith in this region. Given that the original materials differed in colour (because of slight compositional differences, for example), slight differences in colour would be expected to persist in material derived from the original by impact churning.

Mounds in craters may be produced by the updoming of the floor of a crater following a high-velocity impact, as OBERBECK and QUAIDE (1967) have shown; and blocks, which may be identified with the missile, are commonly present in low-velocity impact craters such as those produced by ejecta from explosion craters. The small eminences within the lunar eumorphic craters are regarded as inert missiles for reasons given elsewhere (FIELDER and FIELDER, 1968). Blocks are not visible in the craters in the flow simply because, it is suggested, they plummeted through a more readily compressible material (highly porous lava) than that

Fig.11. Probable lava-filled ring, identified as *P1* in Fig.4. (NASA photo.)

of the substratum of the older unit f1. This explanation is consistent with the anomalously great depths of the inner craters of the double objects on the flow.

In summary, the double eumorphic craters direct attention to two important possibilities: (*1*) that the original surfaces of both the units f1 and f3 have been modified to a depth of several metres; and (*2*) that the more recent unit in the region is the more compressible. Whatever the origin of the surface layer, its presence

may well account for the lack of flow ridging on the most recent flow f3.

The origin of other craters in the region

The excess of craters larger than about 200 m on f3 cannot be accounted for in terms of the differing strength, and response to impact, of the substratum of lava compared with that of unit f1 if the more recent lava of f3 is more crushable than the basal material of f1; in such a case primary (or secondary) impacts would generate fewer non-double craters larger than ~ 100 m in diameter in the flow material. (Impact craters of this size would normally be deep enough to penetrate the surficial layer of debris.) Just the reverse situation arises.

Now the majority of these craters in the areas *H1* and *H2* (see Fig.4) are without raised rims and may, for this reason, be craters of collapse. Those with slightly raised rims may be denuded impact craters. Some, such as those small, rimmed craters in short chains that trend in the directions of the grid system, may be produced by the explosive release of volatiles in late intrusive rocks; clearly, tectonic fractures cannot have been present in the flow materials until after their induration.

Explosion craters are not common on terrestrial basalt flows. By contrast, collapse craters are common on basaltic lava flows on Earth; and the statistical evidence (FIELDER et al., 1971) for endogenic craters—probably craters due to outgassing and collapse—on the lunar flow f3, together with the evidence for collapsed lava tubes, adds weight to the hypothesis that the flow consists of low-strength material.

The depths of collapse craters in terrestrial flows approximate, or are less than, the thickness of the flows, and the same is true for most of the 225-m or larger craters on flow f3. Thus, submorphic craters on the Moon are not necessarily denuded impact craters; a high proportion of them may have been in the submorphic class from the beginning, although the actual proportion of them may vary from one part of the Moon to another. The identical conclusion has been drawn by G. Simmons et al. (personal communication, 1969).

In all probability there is, on units f1 and f2, a mixture of craters of explosive (predominantly primary impact) origin, of secondary impact origin, and of endogenic (predominantly collapse) origin; and the endogenic craters in f3 outnumber those in f1 because of differences in the character and strength of the materials. Below a critical size, however, collapse craters would not form in a lunar lava because the material would have sufficient strength to support the weight of the lava roof.

The process of collapse under gravity may not be the only mechanism responsible for the formation of most of the endogenic craters; the escape of gas (contained by the melt, for example) and the undermining of the porous surface by liquid lava may have assisted the collapse process. In low-strength materials such collapse craters are likely to be circular, and the fact that fresh lunar lavas may be considerably more porous and, therefore, of lower strength than terrestrial lavas of the same composition leads to the prediction that collapse craters in the lunar lava flows are likely to be more circular than those of the same size in terrestrial flows.

Craters smaller than the critical size, 200 m or so, may be thought of as being predominantly primary impact craters. All units are liberally peppered with them, and they have equal number densities in the units f1 and f3. Other small craters may be the result of late drainage of the regolith material.

Two conspicuous volcano-like structures occur in the unit f1, northwest of the principal flow front of f3, and may be seen in Fig.2. One, *R1* in Fig.4, 29 km from the front, is about 2 km in diameter, has a concentric pattern near the inner rim crest, and a floor that is apparently not lower than the surrounding mare. It looks like a lava and tuff ring. The other, *P1* in Fig.4, appears (Fig.11) to be a lava plateau about 1.3 km in diameter and is only a few kilometres from the principal flow front. A third circular plateau structure, *P2*, occurs in a flow unit shown in Fig.4.

The structures *P1* and *P2*, which are of similar size, are covered by the high-resolution photography. Each plateau carries a striking group of craters. Two of the sharpest craters in the group on *P2* have raised rims and their inner walls and floors are strewn with blocks. The largest crater in the group on *P1* has a very low rim and few rocks inside it. At least two dimple (drainage) craters (s in Fig.11) are present on *P1* and at least one on *P2*. Their small size demonstrates again that the regolith is of low cohesion and is capable of drainage in such a way as to form symmetric craters. Both plateaus are respectively surrounded by a zone, of diameter 1.5–2 times that of the plateau, which has a "washed out" appearance. There are relatively very few craters in this zone. Possibly lava spilled over the rim of a crater filled with lava, in each case, and covered the smaller craters which may be presumed to have existed in the immediately surrounding terrain.

Possibly ash from late eruptive craters in the plateau was showered in the vicinity and produced the observed blanketing. In any event, the structures *P1* and *P2* are undoubtedly volcanic and the implication is that even some of the larger craters counted in this study may be volcanic too. It is hardly surprising that one should find at least some volcanic craters in a region of extensive lava flows.

References

ARTHUR, D. W. G., AGNIERAY, A. P., HORVATH, R. A., WOOD, C. A. and CHAPMAN, C. R., 1964. *Commun. Lunar Planetary Lab. Univ. Arizona*, 3: 1.

BALDWIN, R. B., 1949. *The Face of the Moon*. University of Chicago Press, Chicago, Ill.

FIELDER, G., 1961. *Structure of the Moon's Surface*. Pergamon, London.

FIELDER, G., 1965. *Lunar Geology*. Lutterworth, London.

FIELDER, G. and FIELDER, J., 1968. *Boeing Sci. Res. Lab. Doc.*, D1-82-0749.

FIELDER, G., GUEST, J. E., WILSON, L. and ROGERS, P. S., 1967. *Planetary Space Sci.*, 15: 1653.

FIELDER, G., FRYER, R. J., TITULAER, C., HERRING, A. and WISE, B., 1971. *Phil. Trans. Roy. Soc.* (In press.)

KUIPER, G. P., 1965. *Jet Prop. Lab., Tech. Rept.*, 32-700: 9.

MULLER, P. M. and SJOGREN, W. L., 1968. *Science*, 161: 680.

OBERBECK, V. R. and QUAIDE, W. L., 1967. *J. Geophys. Res.*, 72: 4697.

OBERBECK, V. R. and QUAIDE, W. L., 1968. *J. Geophys. Res.*, 73: 5247.

O'HARA, M. J., BIGGAR, G. M. and RICHARDSON, S. W., 1970. *Science*, 167: 605.

STROM, R. G. and FIELDER, G., 1968. *Nature*, 217: 611.

UREY, H. C., 1952. *The Planets*. Oxford University Press, Oxford.

3

Sinuous rilles

J. B. MURRAY

General properties of sinuous rilles

PICKERING (1904) was the first to note the distinctive characteristics of sinuous rilles and to draw attention to them as a separate class of rille. Commonly sinuous rilles commence in a crateriform depression (the head) and decrease systematically in width and depth as the distal end (the tail) is approached. Some of the rilles experience "rejuvenation", becoming suddenly deeper or wider. Others broaden towards the tail as the depth tends to zero. Invariably, the head is topographically higher than the tail of a sinuous rille and the trend of the rille is generally downhill.

There are several cases of sinuous rilles with distributaries and re-entrant loops; and there are some cases in which two sinuous rilles depart from the same crateriform depression in different directions. Discontinuities are found in many sinuous rilles; with some the onset is gradual, but with others the rille takes the form of a series of elongate depressions (see especially the stretches marked SS in Fig.1 and Fig.13, see p.37). Portions of some sinuous rilles (Fig.1) are relatively straight and these portions appear to be indistinguishable from normal rilles as defined by FIELDER (1961). In four cases, one side of the linear stretch of a sinuous rille is higher than the other side, suggestive of a fault.

In five cases, a narrower rille occurs within a wider one, usually emerging from a small depression within the crater at the head of the larger rille, as in the case of Schroeter's Valley (Fig.2). Most sinuous rilles avoid the summits of ridges and show a clear tendency to be deflected by topographic highs (G in Fig.3 and Fig.14, see p.37).

Distribution of sinuous rilles

A map (Fig.4) showing the locations of sinuous rilles has been prepared from high-resolution Orbiter 4 photographs supplemented by medium-resolution Orbiter 5 coverage. Some 85 % of the rilles lie in the maria. Sinuous rilles show a distinct concentration around the edges of Mare Imbrium and Mare Orientale, and there is a suggestion of concentrations around the peripheral regions of other circular maria (HARTMANN and KUIPER, 1962). The main exceptions to this rule are found in the areas of the Marius Hills and the Aristarchus Plateau, where the number-densities of sinuous rilles exceed those anywhere else on the Moon.

The Marius Hills (MCCAULEY, 1967; see also Chapter 4) is an area of well established volcanism with fine examples of pyroclastic cones, lava flows and domes, calderas and fissure eruptions. There is also evidence for volcanism in the vicinity of the Aristarchus Plateau in the dome Herodotus Ω, the probable volcanic fissure eruptions (F in Fig.5), and a possible caldera within Aristarchus F.

Again, transient phenomena have been reported (GREENACRE and BARR, 1964; MIDDLEHURST, 1967; MIDDLEHURST et al., 1968) in and around Aristarchus, at the head of the Herodotus rille, and in the peripheral areas of the circular maria, where probable eruptive phenomena (Chapter 4) and lunar domes also tend to be concentrated.

Some 53 % of sinuous rilles are close to apparent volcanic collapse craters, pyroclastic cones, calderas, domes, flows, the products of fissure eruptions, fissures, and lava lakes. In short, sinuous rilles are found to be linked with areas where there is a record of past volcanicity.

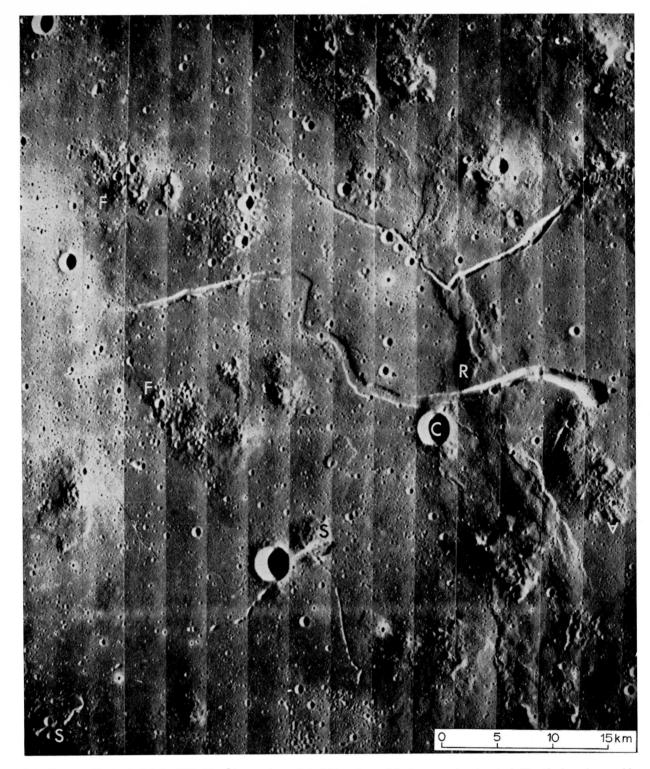

Fig.1. Sinuous rilles in the Marius Hills. Note the angular and straight sections of the two prominent central rilles, the broad mare ridge that lies beneath them, and the crater with its rim unit covering part of the wider rille. Note also the cratered sinuous rille in the lower left part of the frame with long discontinuities in it. (Orbiter 5 photo, NASA.)

0   10   20km

Fig.2. The head of Schroeter's Valley. (Orbiter 5 photo, NASA.)

The analogy with river beds

PICKERING (1904) was the first to suggest that sinuous rilles were dry river beds, and the idea that sinuous rilles owed their origin to the flow of water was revived by FIRSOFF (1959), UREY (1967), LINGENFELTER et al. (1968), and PEALE et al. (1968). The main evidence for the water theory is in the resemblance of the meanders of some sinuous rilles to those of mature river channels.

However, sinuous rilles do not exhibit a drainage basin or a dendritic pattern of tributaries. Generally, they show no broadening as they travel down a slope. Hence they cannot have been produced by an evaporation–rainfall cycle. It is true that certain terrestrial inland rivers degenerate and vanish in desert areas but, unlike rilles, inland rivers have a network of tributaries in their upper reaches.

The form of the channel of a sinuous rille differs considerably from that of a river channel on Earth: the lunar channel is an order of magnitude deeper in relation to its width than the terrestrial river channel (Fig.6); and the ratio L/b, where L is the "wavelength" of meanders and b the channel width, is about one-fifth that pertaining to terrestrial rivers (Fig.7). Furthermore, sinuous rilles exhibit none of the characteristic deeps on the outer, leading edge of a meander, nor do the rilles have lower sand-barred terrain on the inside. The commencement in a crateriform depression is quite uncharacteristic of terrestrial rivers.

In water channels (P. Ackers and F. G. Charlton, personal communication, 1970), meanders develop as

J. B. MURRAY

Fig.3. Hadley rille. Note the diversion of the rille by the mountain (top right) and the partial covering of the rille by deposits associated with the crater (centre). (Orbiter 5 photo, NASA.)

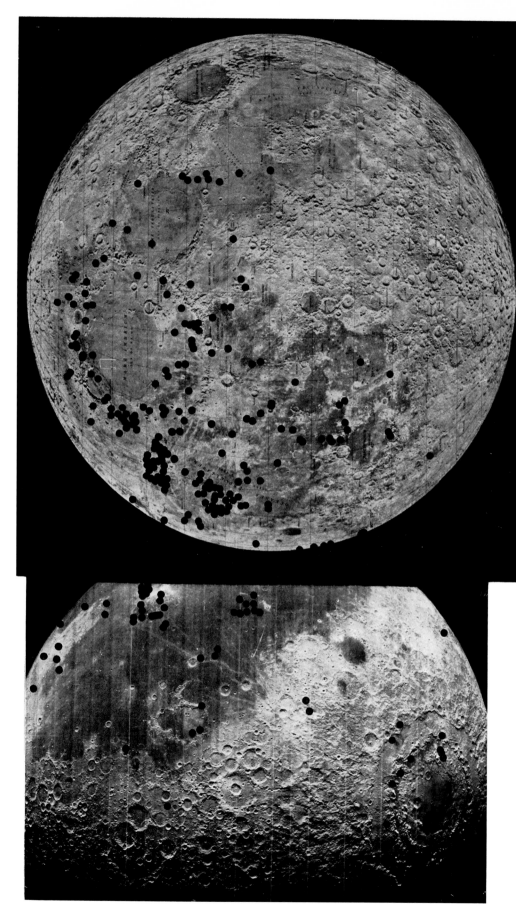

Fig.4. Maps of the distribution of sinuous rilles, the left photograph showing those on the western limb of the Moon. The right hand map may be incomplete in the eastern (right-hand) part, as many of the Orbiter 4 frames in this region are degraded. The black spots represent the position of the source crater or, if there is no source crater, the centre point in the track of the rille. (NASA.)

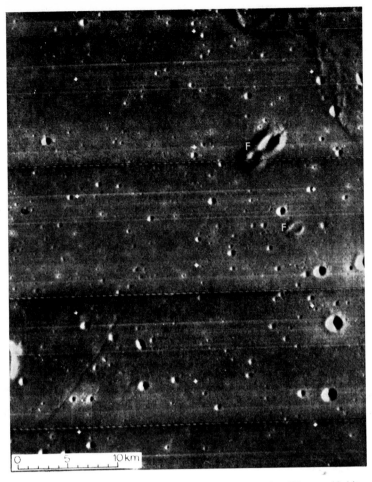

Fig.5. Apparent volcanic fissure eruption craters northwest of the Aristarchus Plateau. (Orbiter 4 photo, NASA.)

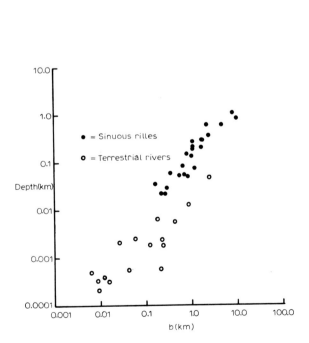

Fig.6. Channel depth versus channel width, b, for terrestrial rivers and sinuous rilles.

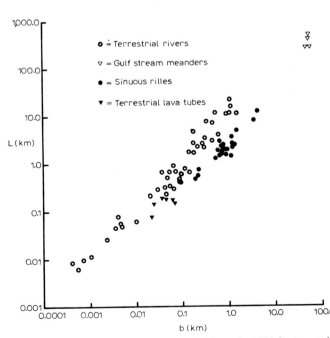

Fig.7. L (meander "wavelength") versus b (channel width) for terrestrial rivers, sinuous rilles, terrestrial lava tubes and meanders in the Gulf Stream.

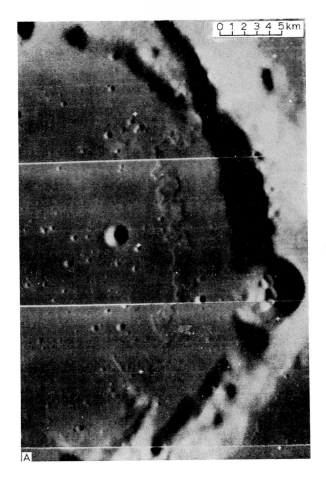
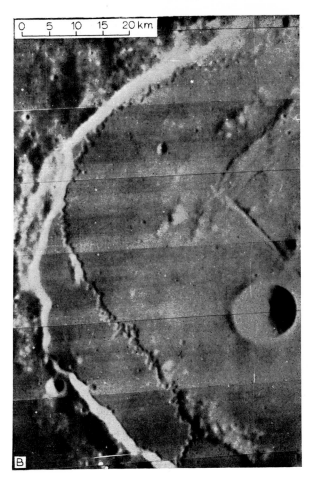

Fig.8. Sinuous rilles in A: Ulugh Beigh A; and B: Posidonius. Note the extreme sinuosity of the channels and the cut-offs. (Orbiter 4 photos, NASA.)

a result of the movement of sediment along the bed of the channel and the resulting deposition of shoals at the sides. The shoals grow until they are of such a size that they deflect the main stream of water towards the side of the bank that is concave inwards, thus increasing the scouring there. Meanders develop, the ratio r_m/b, where r_m is the radius of curvature of the meander, tending to a value between 2 and 3. The reason for this tendency was discussed by BAGNOLD (1960), who demonstrated theoretically that the resistance to flow in a pipe or channel of uniform cross section fell to a sharply defined minimum within the narrow range of approximately 2–3 of the ratio r_m/b.

In the laboratory experiments of P. Ackers and F. G. Charlton, streams 1 m wide and 100 m long took two to three weeks to develop even semi-mature meanders. Larger streams take progressively longer to form meanders. Mature meanders, such as those in Fig.8, would probably take tens or hundreds of years to form, were they on the Earth, even with the river in the critical bankfull condition. Calculations based on the formulae of RUBEY (1933) and COLBY (1961) indicate

that an unrealistically large subsurface volume of water would be required in the Moon if the very mature meanders of the sinuous rilles were to have been formed by the flow of water across the lunar surface.

It should also be noted that, in small-scale experiments conducted in vacuo, ADLER and SALISBURY (1969) injected water in the surface of a bed of particulate material; this failed to produce any channels whatsoever. It seems that sinuous rilles cannot possibly be the result of erosion, transport and deposition of sediment by flowing water.

The analogy with lava tubes

The abrupt discontinuities in a sinuous rille, and the presence of a crateriform depression at its head, are characteristics similar to those of some terrestrial lava tubes with partially collapsed roofs (see Fig.9). KUIPER et al. (1966) drew attention to this fact.

South of Sinus Iridum there is a cratered, sinuous

Fig.9. Collapsed lava tube in Idaho. Note similar, shallower feature on the left. (U.S. Air Force photo.)

ridge (Fig.10) which starts in a sinuous channel, but changes into a sequence of collapse craters, craters with raised rims, and ridges. OBERBECK et al. (1969) argued convincingly that this was a partially collapsed lava tube similar to the feature originating at Modoc Crater (California). The lunar feature is an order of magnitude larger than the terrestrial example, but calculations show that lava roofs as wide as those indicated in Fig.10 can be supported under lunar gravity.

The depth/width ratio of the terrestrial lava channel depicted in Fig.9 is about the same as that characterising sinuous rilles; the channel in Fig.9 has many re-entrants and other neighbouring, associated collapse depressions (lettered C) such as are found in association with many of the lunar rilles (Fig.11). The roof of the terrestrial lava tube has slumped differentially, to give the variations in width and depth characteristic of some sinuous rilles. Both the width and the depth of the terrestrial feature decrease as the distance, measured along the channel from the source crater, increases— and this is also true in the case of the sinuous rilles. The second, shallower crater (L in Fig.9) and its channel, to the left of the more prominent crater, is paralleled by the case shown in Fig.11.

Fig.12 shows two separate, collapsed lava tubes in Iceland, emerging from one crater. A parallel case (Fig.13) occurs southeast of the lunar crater Marius.

The neighbouring lava field (Fig.12) is again riddled with collapse depressions, some showing a tendency to alignment. Note that one channel has developed two prominent, meandering sections (M and G), one of them (G) a "gooseneck" with a re-entrant or distributary, as happens in many of the goosenecks in the crater Ulugh Beigh A (Fig.8A). It is also significant that the meanders of the lava streams do not exhibit "deeps" near to the outer bank on a bend, and that the inner bank does not consist of low "shoals".

The crateriform depression at the heads of the terrestrial lava tubes shown in Fig.12 is elongated, a trait that characterises the heads of many lunar sinuous rilles. As in the previous example, the width and depth of the channel in Fig.12 decreases, though with occasional interruptions, from the source crater to the distal end. The branch labelled MG terminates in a fissure in the lava flow. Some of the sinuous rilles on the Moon terminate in fractures.

The processes by which a lava tube develops meanders are significantly different from the processes operating in a terrestrial water channel. LEOPOLD and WOLMAN's (1960) compilations show that, in natural and artificial open water channels, the ratio r_m/b remains more or less constant whatever the size of the channel. Meanders of lava tubes are formed in the time taken for the lava to solidify. When the surface of a lava

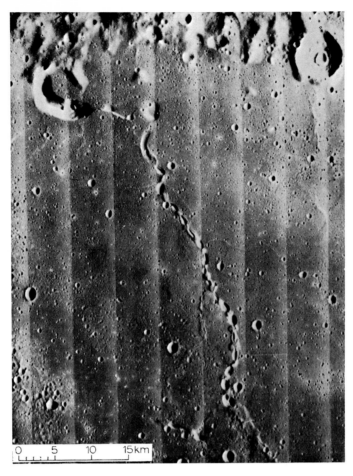

Fig.10. A complex feature, southwest of Sinus Iridum, composed of sinuous channels, raised hills, raised cones, and rimless depressions.
(Orbiter 5 photo, NASA.)

stream solidifies it may enclose the flowing lava, and later collapse of the lava roof may reveal the course of the channel beneath. The ratios r_m/b and L/b (Fig.7) are generated without recourse to the erosion and deposition mechanisms of water streams; so the final form of an ideal lava channel is a maturely meandering feature with no steep outer bank associated with the meander and no cuspate form of the inner bank.

A parallel to this has been observed in the Gulf Stream (LEOPOLD et al., 1964), which consists of one liquid flowing through another of slightly different viscosity, and which exhibits giant meanders having the values r_m/b and L/b similar to those of lava streams.

The sequence of events described above holds for lavas of less than a certain critical viscosity. When the viscosity of the lava exceeds this critical value the resulting channels will be of an immature meandering, or non-meandering, character. This character may typify all, or only part, of the channel. The meanders of the channel in Fig.9, for example, seem to be rather underdeveloped, particularly near to the crater. The means of r_m/b and L/b are also found to vary from one

stretch to another of the same sinuous rille.

Some 37% of sinuous rilles have no crater at the head. Evidently, many of them represent portions of key-type sinuous rilles which have been interrupted— for stretches of up to 100 km— by later lava flooding. In such cases, the two or more parts of the rille either display similar cross-sections, depths, widths, r_m/b ratios and sinuosities[1], or, if these parameters alter in a fairly regular fashion along one part of the rille, the trend is continued in the other part of the rille. Later lava flooding may introduce the irregular variations in depth or cross-sectional area found in some sinuous rilles.

A few sinuous rilles pass over hills. Fig.11 shows a rille in the Marius Hills, that intersects a wrinkle ridge R. If folding or up-arching of the surface occurred beneath the rille, and on a large scale compared with the width of the rille itself, it could have been raised

---

[1] The sinuosity of a terrestrial river is defined as the ratio (channel length)/(valley length). Since sinuous rilles generally have no valleys, sinuosity is taken to be the ratio (length of rille)/(length of rille course, ignoring meanders).

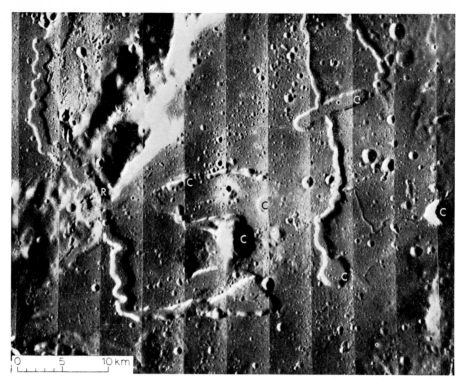

Fig.11. Rilles near the Harbinger Mountains. Note the rille passing through a mountain range *(R)*, and the many apparent collapse depressions *(C)* in the region. (Orbiter 5 photo, NASA.)

Fig.12. Two collapsed lava tubes linking with a single crater in Iceland. (U.S. Air Force photo.)

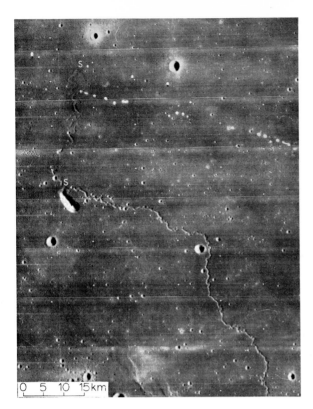

Fig.13. Two sinuous rilles, the northern one discontinuous, associated with a single, elongated depression. (Orbiter 4 photo, NASA.)

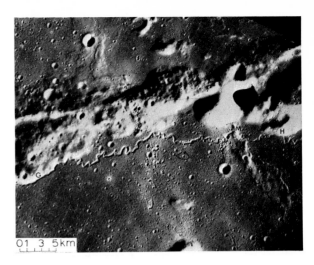

Fig.14. Sinuous rille just northwest of the Aristarchus Plateau. The rille disappears beneath highland (right) and a mare ridge; it hugs the edge of the high ground on the left. (Orbiter 5 photo, NASA.)

up without interruption. On the other hand, the ridge *M* in Fig.14 interrupts a rille.

Again, Rima Prinz II (Fig.11) intersects the Harbinger Mountains at *R*, and the channel is twice as deep before passing through the mountains. One possibility is that the lava was ponded up by the mountains and flowed through the lowest pass when it reached the appropriate level. The angle of slope of a lava flow will probably affect the sinuosity of its channel. It is interesting to note that the most highly sinuous channels found on the Moon are those inside Ulugh Beigh A (Fig.8A) and Posidonius (Fig.8B). In both cases the rilles are inside large craters, where the surfaces of the lavas slope very little.

Schroeter's Valley and its inner rille (Fig.2) exemplify highland and mare types of sinuous rilles. The valley itself cuts highland material, and the sinuosity is 1.1. However, the smaller rille has the high sinuosity of 2.5 in the mare-type floor of the valley and, where it breaks out of the valley and flows through highland material, the sinuosity drops to 1.2—about the same as for Schroeter's Valley itself.

The highly sinuous bends of the inner rille frequently disappear into the banks of Schroeter's Valley and reappear again. This is presumably due to the banks of Schroeter's Valley having become wider and less steep, under the action of the lunar erosive processes, and having encroached upon the rille.

These facts argue heavily in favour of a lava tube origin for many of the lunar sinuous rilles. However, it is important to note certain differences between terrestrial lava channels and lunar sinuous rilles:

(*1*) The size of the lunar channels is an order of magnitude greater than that of the terrestrial examples.

(*2*) The very high sinuosities of the rilles shown in Fig.8, and a few others, are not found in lava tubes or channels on Earth.

(*3*) Many terrestrial lava tubes are situated in the highest part of a flow, where the lavas are thickest.

(*4*) One can rarely[1] see lava fronts associated with the flows in which the sinuous rilles are supposed to have occurred.

All these differences may be explained readily in terms of the lower viscosity of lunar lavas, discussed by FIELDER and FIELDER (1968; see also Chapter 2) and MURASE and McBIRNEY (1970).

Ages and stratigraphic relations

The ages of many sinuous rilles may be established relative to the ages of neighbouring features. There are a number of cases of wrinkle ridges (*M* in Fig.14) and

---

[1] R. G. Strom (personal communication, 1970) has pointed out that the sinuous rille near to the Tobias Mayer domes is associated with a flow front which appears to have been arrested by one of the domes. There also seems to be a flow front (*FF* in Fig.1) associated with the largest rille shown there.

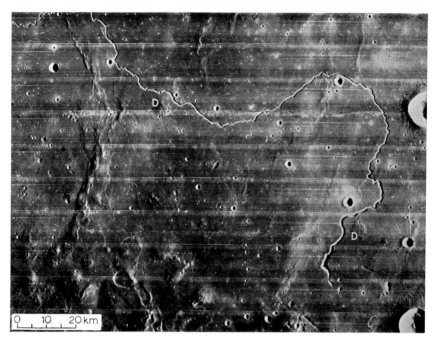

Fig.15. Sinuous rille deflected by two domes in an area rich in volcanic domes and calderas (lower left). (Orbiter 4 photo, NASA.)

of the rim materials of craters (C in Fig.1 and 3) interrupting rilles, the rilles in these cases apparently being formed before the craters or ridges. On the other hand, there are many cases of sinuous rilles being diverted by craters or other topographic highs (D in Fig.15), and of sinuous rilles hugging the edge of high ground (G in Fig.3 and 14), indicating that the craters and topographic highs were formed before the sinuous rilles. An instance of higher material actually encroaching upon a rille and covering it has been cited in Schroeter's Valley; another example of this phenomenon is the long rille northwest of the Aristarchus Plateau (Fig.14), which appears to be younger than the mountain range H, but older than the scree-like formation that covers one of the meanders.

All sinuous rilles in the Aristarchus region (Fig.2 and 14) are overlain by rim materials or secondary craters from Aristarchus indicating that the rilles are older than the crater. Similarly, some sixteen sinuous rilles in the vicinity of Copernicus have rim materials or secondary craters superimposed on them; and the sinuous rilles near to Kepler, Bullialdus, Aristillus, Aristoteles and Eudoxus are all similarly affected. In every case large craters of the Copernican Era postdate sinuous rilles.

Conversely, other sinuous rilles appear to be later than the deposits around craters of the Archimedian Series, such as Plato, Herodotus, Krieger and Sinus Iridum. Thus, nearly all sinuous rilles appear to be contemporaneous with mare material of the Procellarum Group.

In general, sinuous rilles are to be dated with the late stage lavas of the circular maria. The Aristarchus Plateau and the Marius Hills are centres of lava emission apparently unconnected with the formation of the circular maria but of about the same age.

References

ADLER, J. E. M. and SALISBURY, J. W., 1969. *Science*, 164: 589.
BAGNOLD, R. A., 1960. *U.S., Geol. Surv., Profess. Papers*, 282-E: 135.
COLBY, B. R., 1961. *U.S., Geol. Surv., Water Supply Papers*, 1498-D: 1.
FIELDER, G., 1961. *Structure of the Moon's Surface*. Pergamon, London, p.198.
FIELDER, G. and FIELDER, J., 1968. *Boeing Sci. Res. Lab. Doc.*, D1-82-0749: 1.
FIRSOFF, V. A., 1959. *Strange World of the Moon*. Hutchinson, London, p.159.

GREENACRE, J. and BARR, E., 1964. *Aeronaut. Chart Inform. Center, Tech. Papers*, 12: 1.
HARTMANN, W. K. and KUIPER, G. P., 1962. *Commun. Lunar Planetary Lab. Univ. Arizona*, 1: 51.
KUIPER, G. P., STROM, R. G. and LE POOLE, R. S., 1966. *Jet Prop. Lab., Tech. Rept.*, 32-800: 199.
LEOPOLD, L. B. and WOLMAN, M. G., 1960. *Bull. Geol. Soc. Am.*, 71: 769.
LEOPOLD, L. B., WOLMAN, M. G. and MILLER, J. P., 1964. *Fluvial Processes in Geomorphology*. Freeman, London, p.296.

LINGENFELTER, R. E., PEALE, S. J. and SCHUBERT, G., 1968. *Science*, 161: 266.

McCAULEY, J. G., 1967. *U.S., Geol. Surv., Map I-491*.

MIDDLEHURST, B. M., 1967. *Rev. Geophys.*, 5: 173.

MIDDLEHURST, B. M., BURLEY, J. M., MOORE, P. and WELTHER, B. L., 1968. *NASA, Tech. Rept.*, TR R-277.

MURASE, T. and McBIRNEY, A. R., 1970. *Science*, 167: 1491.

OBERBECK, V. R., QUAIDE, W. L. and GREELEY, R., 1969. *Mod. Geol.*, 1: 75.

PEALE, S. J., SCHUBERT, G. and LINGENFELTER, R. E., 1968. *Nature*, 220: 1222.

PICKERING, W. H., 1904. *The Moon*. Doubleday and Page, New York, N.Y., p.42.

RUBEY, W. W., 1933. *Trans. Am. Geophys. Union*, 14: 497.

UREY, H. C., 1967. *Nature*, 216: 1094.

4

Centres of igneous activity in the maria

J. E. GUEST

Introduction

Dark material fills, or partially fills, the marial basins. It is clear from the way this material enters embayments on the margins of the surrounding highlands that it is generally stratigraphically younger than highland material. With an impact origin for the majority of lunar craters, the relative youth of mare material is further substantiated by its having a much lower crater density than highland areas. The dark mare fill has been mapped as the Procellarum Group (McCAULEY, 1967a). The lack of any satisfactory sedimentary process on the Moon for filling mare basins leads to the view that the relatively level-surfaced mare units result from volcanic activity. The depth of burial of crater rings appears to differ from one unit to another: this suggests that the mare units were emplaced to various thicknesses in their respective basins.

These views find support in the presence on many marial surfaces of features analogous to those observed in volcanic regions on Earth. The most striking of these lunar features are the extensive flows (Chapter 2) in Mare Imbrium. Analyses carried out by Surveyor soft landing craft showed that the composition of mare material was consistent with a basaltic composition; and rock returned from Apollo missions proved that mare material at the sites visited was of igneous origin.

One of the problems relating to these regions is the scarcity of source vents to produce marial units of such large volume. Clusters of volcanic vents occur locally but, over vast areas of the maria, no such vents may be resolved. There are several possible explanations for this. It is likely that many units were erupted rapidly as low-viscosity lavas with a low gas content. Lavas erupted with little explosive emission of gas would not form pyroclastic cones of any significance over the vent and would leave little trace of the source, unless wholesale caldera collapse were to accompany or follow the event. Low-viscosity lunar flows would travel considerable distances from their sources and would, in any case, be difficult to trace back to their original source. Also, evidence of source areas has probably been removed by denudation. None of the marial areas is young enough to have avoided peppering by impact craters, the debris from which has built up a fragmental regolith. This layer varies in thickness depending, presumably, on the age of the surface. In areas where the regolith is thick (say 15–20 m) the terrain will have been partially levelled and covered by the regolith, and small-scale surface features of flows, together with source vents, will tend to be removed.

Despite the apparent scarcity of source vents for many marial flows there are sufficient vents to demonstrate that not only are there volcanic vents of types analogous to those found on Earth, but that there is a diversity of lunar volcanic features. In certain parts of the Moon there are well marked igneous centres consisting of clusters of volcanic vents either all of the same type or consisting of closely spaced vents of different types. The centres usually relate to discrete rock stratigraphic units in the maria and do not provide source vents for the lavas that are widespread over the marial plains.

Returned igneous rocks

Of all the evidence for an igneous origin of the mare material, that pertaining to the rocks returned by Apollo missions is the most conclusive. All the rocks from Mare Tranquillitatis[1] were collected from

---

[1] Suggested reading: *Science*, 167: 447 et seq. (1970).

the regolith, which consisted of crystalline rocks enclosed in a fine, fragmental matrix. The crystalline rock fragments are considered to be from the underlying bedrock, and most of the matrix is thought to have been produced by fragmentation and impact melting of the bedrock.

The crystalline rocks are known to be igneous because they contain pyrogenic mineral assemblages suggestive of crystallisation from a melt. Mineral species identified in the crystalline rocks—mainly ilmenite ($FeTiO_3$), plagioclase feldspar ($CaSi_2Al_2O_8$, with about 15% Na substituting for Ca) and clinopyroxene (Ca, Mg, Fe silicate)—are those normally found in terrestrial igneous rocks of basaltic composition. The chemistry (Chapter 10) of the rocks shows them to differ markedly from the majority of terrestrial rocks, the main differences being the high Ti and Fe and low alkali contents. These differences in chemistry are reflected in the modal composition, there being an unusually high percentage of ilmenite in the rocks. High concentrations of Ti and Fe are rare in terrestrial rocks but, where they do occur, they are normally accompanied by high amounts of alkalies.

Mineral assemblages in the lunar rocks indicate that crystallisation could only have taken place under low partial pressures of oxygen; experimental studies suggest that crystallisation took place at temperatures of about $1,200-1,500\,°C$. During crystallisation, cooling probably took place relatively rapidly, as pyroxene crystals show wide variations in composition from Mg-rich cores to Fe-rich margins. Zoning of this type indicates the metastable character of the minerals which must have crystallised too rapidly for them to have remained in stable equilibrium with the melt. Sub-microscopic intergrowths of Ca-rich and Ca-poor pyroxene also suggest rapid cooling at high temperature.

The most obvious textural similarity between rocks from Mare Tranquillitatis and terrestrial volcanic rocks is the presence in the lunar rocks of vesicles, indicating some de-gassing during eruption. It is not clear, however, how much gas was present at the time of the lunar eruptions: the lack of hydrous minerals would suggest that water was absent at the time these lavas were erupted on the Moon. Of considerable importance to the study of lunar petrogenesis, and also in the interpretation of possible lunar volcanic morphologies, is the presence, in these rocks, of interstitial material of acidic composition. This suggests that lunar magmas are potentially capable of generating magmas of more acidic composition by fractionation. It is thus unlikely that all the marial areas consist of lavas of the same composition as those returned from

Mare Tranquillitatis; and, furthermore, features indicative of the eruption of lavas with different properties may be expected.

Analyses of rocks collected by the Apollo 12 astronauts in Oceanus Procellarum resemble those from Mare Tranquillitatis, and confirm an igneous origin for mare material[1]. They also confirm the view that differentiation has occurred to give rocks of different compositions, for, in these rocks, there is a much wider range of rock type, both in texture and modal composition, than in the Mare Tranquillitatis rocks: SiO_2 contents, for example, range from 36% to 61%. Such a range in composition clearly points to the conclusion that volcanic activity of a variety of types must have occurred.

Work by MURASE and McBIRNEY (1970) on synthetic melts of the same composition as the Apollo 11 rocks has shown that lavas of this composition could have viscosities of less than 10 poises at normal eruption temperatures. Such low viscosities may be related to the high iron and titanium contents together with low silica. Few terrestrial lava flows have viscosities as low as this, and are usually at least an order of magnitude more viscous. According to MURASE and McBIRNEY (1970) lunar lavas of this composition would travel at about twice the velocity of a basalt on Earth and could cover much wider areas than those covered by most terrestrial lavas. This would explain the vast surface areas of the flows in Mare Imbrium (Chapter 2). Again, it is now easy to see how some mare flows may have only low flow fronts. Murase and McBirney have also argued that lava tubes larger than those normally found on Earth could develop in lunar lavas of this composition. Their experiments have shown that low-viscosity and, by implication, highdiffusion-coefficients, allowed the growth of the large crystals present in the Apollo samples.

By contrast with the emission of lavas in the maria, there is some evidence for explosive volcanism at the Apollo 12 site; the astronauts returned samples of a layer in the regolith of what may be described as volcanic pumiceous ash.

Types of source vent in marial regions

A number of types of feature mark the vents from which lunabase was erupted. These types will be described in this section.

---

[1] Suggested reading: *Science*, 167: 1325 et seq. (1970).

Lunar domes

These are generally low, smooth-surfaced features with convex slopes of up to 2° or 3°. The diameters of the domes range from a few kilometres to about 10 km. Two of the Cauchy domes, with diameters of about 10 km, have respective heights of 100 m and 49 m (RIFAAT, 1967) above the surrounding mare surface. Many domes have terminal crater pits: a crater on one of the Cauchy domes has a depth of about 50 m, according to Rifaat. Domes of this type generally have albedoes similar to those of the adjacent mare plains and are probably composed of similar rock types. These domes may well be analogous to terrestrial shield volcanoes, although SALISBURY (1961) has proposed that they are formed by igneous uparching of surface strata.

A second type of lunar dome has a concave slope-profile with maximum slopes of up to 7°. These domes have less smooth surfaces and are commonly pitted by rimless craters. Some of the domes of this type have no direct analogies on Earth, but are considered by McCAULEY (1968) to have been formed by extrusion of viscous magmas. Others resemble terrestrial flows of viscous magma and still retain the source vents (Fig.10, see p.52).

Calderas

Large-scale eruption of lavas into the mare basins would, by terrestrial analogy, be expected to be associated with caldera collapse depressions. A variety of collapse features occurs in the maria. Many collapse features have the shape and dimensions of calderas of the Kilauean type on Earth; but others, although clearly of volcano-tectonic origin, differ from known terrestrial structures both in size and shape. Examples of caldera-like depressions are shown in Fig.1. Some have steep inner walls with exposed bedrock but most have low slopes and probably have been markedly modified by denudation and the development of the regolith.

In most cases tectonic control is evident, the walls of the collapse depressions paralleling tectonic trends. Many are long, narrow depressions with parallel sides (for example, 7 in Fig.9, see p.51); these are probably an intermediate form between normal caldera collapse and graben faulting which is also common on the Moon.

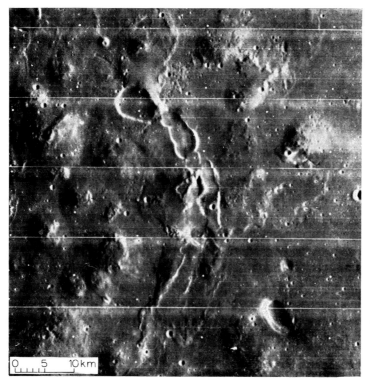

Fig.1. Caldera-like depressions near Marius. (Orbiter 4 photo, NASA.)

Mare ridges

These occur on almost all mare surfaces and are commonly concentric with the margins of circular maria. In almost all cases they appear to be structurally controlled. They have heights of up to ~ 100 m above the surrounding mare surfaces, and widths of up to several kilometres. Although many ridges extend longitudinally for some several hundred kilometres they do not usually consist of a single ridge unit but are broken into shorter ridges which together form a ridge complex. A number of different patterns have been developed within these ridge complexes; most commonly developed are zig-zag, en echelon or parallel patterns (FIELDER and KIANG, 1962). The U.S. Geological Survey maps interpret these as being either anticlines or structurally controlled sites of volcanic extrusions or intrusions. Their close association with features of known tensional origin, such as graben, supports the view of FIELDER (1961) that they are of volcanic origin and consist of lava which has forced its way up along fissures or fissure complexes. This is further supported by the frequent occurrence of flow-like bodies extending from the mare ridges (see Chapter 2, fig.1).

The general lack of well defined source vents for relatively fresh flows such as those in Mare Imbrium (Chapter 2) has led to the conclusion that, in general, mare ridges mark the sites of fissures from which large volumes of lava were erupted.

Elementary rings

"Ghost" rings are generally considered to be old craters which have been flooded by younger mare material leaving only a rim projecting above the present surface. FIELDER (1965) proposed a new theory for the formation of these structures, which he termed elementary rings. He regarded them to be relatively young features formed by extrusion of lavas along ring fractures. This idea was developed by GUEST and FIELDER (1968) who contended that at least some of the so-called "ghost" craters were of this origin; they emphasised, nevertheless, that not all incomplete rings were of this type, and that many appeared to be the walls of buried craters.

Elementary rings were considered by Guest and Fielder to consist either of rings of mare ridges (for example, Lamont) or of incomplete rings of hills having an albedo different from that of the surrounding terrain. It was concluded that structures of the elementary ring type had resulted from the development of ring fractures that followed large-scale eruptions of

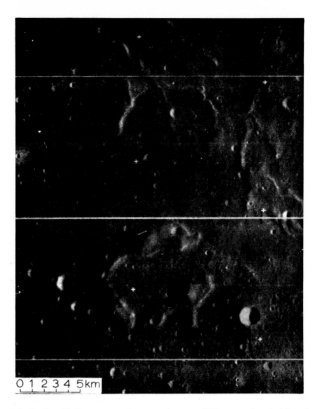

Fig.2. Small elementary ring, southwest of Flamsteed, considered to have formed by the eruption of lava from a ring fracture. (Orbiter 4 photo, NASA.)

volcanic material into the mare basins; extrusion of relatively viscous magma along the fractures generated elementary rings and the form of the rings was controlled by the type of magma erupted. Fig.2 shows a small elementary ring near to Flamsteed.

One difficulty presented by this hypothesis is that of distinguishing between elementary rings and buried craters; many examples could be of either origin. It will be shown in Chapter 6 that Tsiolkovsky has a well defined pattern of faults associated with its rim; this pattern is considered to reflect underlying basement structure which has been re-activated by the Tsiolkovsky event. Similar structural patterns may be observed on all large, fresh craters and around the circular maria such as Mare Imbrium. Impact craters younger than the maria may themselves re-activate faulting, or they may be deformed by later tectonic movements. Mare material is generally only mildly affected by the basement structure except in marial areas surrounding young, large craters where movements initiated by the formation of the crater reflect the basement structure. Where older craters have been flooded by mare material, the crater rims may be expected to have a well developed basement fracture pattern which does not extend into the later mare material. This fault pattern may therefore be used as a stratigraphic guide, and "ghost" craters which have this pattern printed on them, but not on the surrounding mare, may be considered to be older, buried features and not elementary rings.

Dark halo craters

These are circular to elliptical craters surrounded, as the name implies, by a mantle of fragmental material of relatively low albedo. Good examples of this type of crater have been mapped on a scale of 1:250,000 by CARR (1969) on the floor of Alphonsus.

In Alphonsus (Chapter 1, fig.6–9) dark halo craters are located on rilles (graben), indicating structural control and, therefore, endogenic origin; they differ from typical impact craters in having wide rims characteristic of volcanic craters such as maars. Crater diameters range from about 500 m to nearly 3 km on the longest axis; most are elliptical and follow the line of the rille with which they are associated. The dark

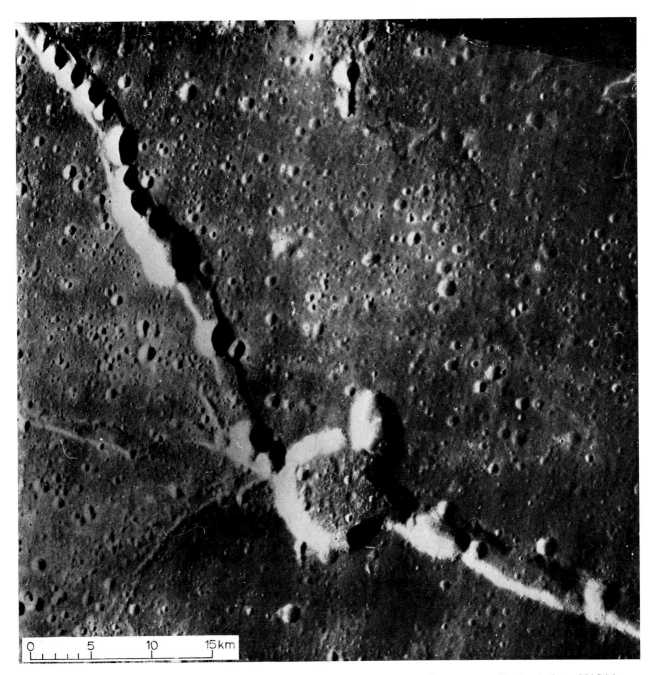

Fig.3. Hyginus rille showing craters formed largely by collapse on an apparent graben structure. (Orbiter 5 photo, NASA.)

deposits extend for as much as 5 km from the rims of the craters, and subdue the surrounding terrain, partly filling adjacent rilles; the way the deposits mantle the surrounding terrain demonstrates that they are fragmental in nature and that the materials are ejecta rather than lava.

Dark halo craters are scattered over the Moon's surface: their distribution has been illustrated by SALISBURY et al. (1967). It appears that these craters are entirely explosive in character and that they are not the vents from which mare material was erupted.

Rille craters have no dark halo but are otherwise similar to dark halo craters. Some rille craters are almost circular. Craters of this class commonly form chains along the rilles; this indicates that the development of rille craters was controlled by the rilles and that the craters are endogenic. Some have wide rims analogous to those of terrestrial maar volcanoes, but others are almost rimless indicative of an origin by collapse. A variety of craters of this type occur on the Hyginus rille (Fig.3) and in the surrounding area (WILHELMS, 1968). Small rille craters are found in Alphonsus. CARR (1969) concluded that these were old, dark halo craters, the halos of which had been destroyed by a process causing mixing with the surface regolith.

Fissure cones

A number of chains of craters, sometimes associated with rilles, occur in marial regions. These resemble typical fissure volcanoes on Earth. Fig.4 shows a photograph of one example of a possible fissure volcano (*F*) on the Moon near to the Flamsteed P ring. Other possible fissure volcanoes are shown in Chapter 3 (fig.5). Clearly, the craters are associated with the rille on which they are situated, and must have been formed by an endogenous process rather than by impact.

Igneous complexes

Although igneous complexes are rather widely scattered over the maria, certain marial units contain complexes consisting of tight clusters of vents. The complexes in Oceanus Procellarum are particularly interesting: three, widely spaced centres are apparently part of a volcanic zone, marked by mare ridges. The main centres on this belt are, from north to south, Rümker, the Aristarchus Plateau, and the Marius Hills. McCAULEY (1968) suggested that this zone of volcanism was analogous to terrestrial mid-oceanic ridges, and that the three centres of volcanism were the product of convection currents localised along the major axis of Oceanus Procellarum. Two of the Oceanus Procellarum igneous complexes are described here.

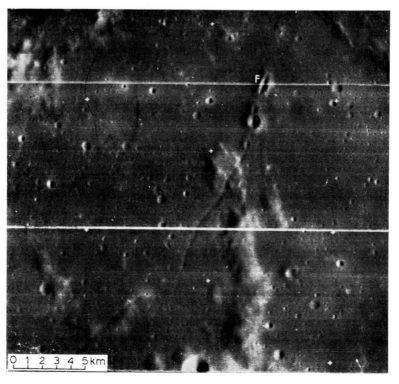

Fig.4. A well-defined fissure to the southwest of Flamsteed. The cones are thought to be formed of pyroclastics erupted from vents on the fissure. (Orbiter 4 photo, NASA.)

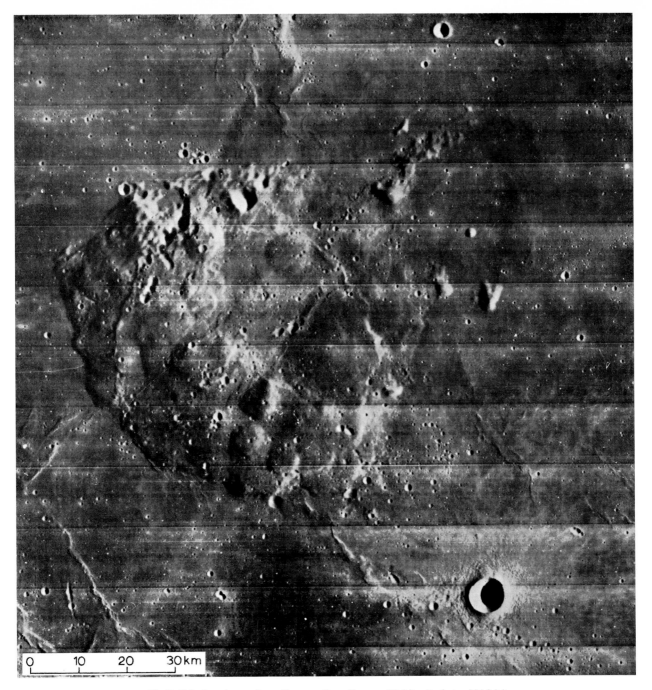

Fig.5. Rümker, in northern Oceanus Procellarum. (Orbiter 4 photo, NASA.)

Rümker

This is a roughly polygonal plateau some 70 km across in northern Oceanus Procellarum at 40 °N 58 °W (Fig.5). It lies on a large mare ridge complex which, in this area, strikes towards the northwest. Isolated remnants of what may be rocks of the Fra Mauro Formation—which is found in some of the regions surrounding Mare Imbrium—project through the near-

by mare surface as inliers. Most of Rümker displays a terrain similar to that of the surrounding mare but differs from the mare by having low domes on the plateau suggesting that this is a centre of igneous activity.

Not all the plateau is necessarily of volcanic origin: the northwestern part of the mass presents a type of terrain that differs from the rest of Rümker in that it has an irregular surface cut by a set of closely spaced, linear valleys striking to the northeast. The terrain is similar to that of much of the Fra Mauro Formation where it has been strongly faulted to give what is often described as the Mare Imbrium "sculpture". On the outer margin of this outcrop mare material has flooded the embayments between ridges of the fracture pattern, and the ridges form "promontories" above the mare surface. On this evidence, and with the observation that the fracture pattern does not continue into the surrounding mare material, it is clear that the northwestern part of Rümker is older than the mare material and that, structurally, this part of Rümker is comparable with outcrops of the Fra Mauro Formation elsewhere. Likewise it may be shown that the mare-like units of Rümker— here called the Rümker Group, for convenience—are younger than the northwest outcrop; but the age relations between the Rümker Group and the surrounding mare material of the Procellarum Group are not so clear from the contact geometry. The surface of the lower units of the Rümker Group are marked by a fine pattern of lineaments parallel to those on the northwest outcrop and follow the local structural trend. This pattern again does not continue —or continues only weakly—across the boundary to the mare surface, and it may be concluded that at least the lower part of this group is mostly older than the Procellarum Group. If this is true then the Rümker Plateau represents an igneous complex which began to form at an early phase of eruption in this mare. Since the lower units of the Rümker Group directly overlie Fra Mauro material in the northwestern part of the Rümker mass, they may represent one of the earliest phases of magmatism in Oceanus Procellarum. Later material of the marial plain possibly interdigitates with units of the Rümker Group. Protracted activity at this site established Rümker as an "island" standing above the mare plain.

The Rümker Group consists of a number of well marked stratigraphic units. Four principal phases in the construction of Rümker are identified on the geologic map (Fig.6). The lowest (unit 1) of these directly overlies the Fra Mauro Formation and outcrops in some places around the edge of Rümker. Unit 1 is best developed on the northern part of the mass. The surface of the unit is smooth and undulating with low, poorly defined domes. The next two units (2 and 3) form an almost circular outcrop with a well defined, outer boundary scarp. In the northwest these units overlap unit 1 to lie directly over the Fra Mauro outcrop. Unit 1 also has a number of low domes the margins of which are defined by slight changes in slope. Unit 3 may be younger than the surrounding mare

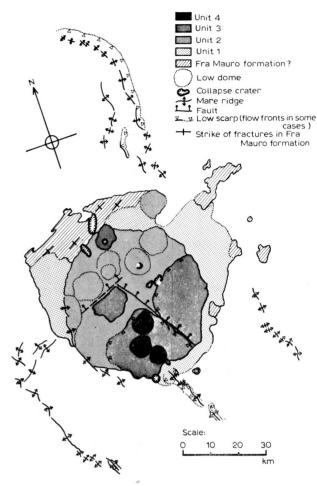

Fig.6. Geologic sketch map of Rümker.

plain. The youngest units (4) consist of domes with steeper, well defined fronts. These domes appear to represent the last phase of volcanism on the plateau.

The Rümker Group appear to have been cut by at least two faults which form faint scarps. These do not extend into the surrounding mare material of the Procellarum Group, again suggesting a greater age for the volcanic material of the plateau.

Numerous rimless craters, presumed to have formed by collapse associated with the volcanic activity, are found in the plateau. Some occur in the Fra Mauro Formation near the Rümker Group boundary, while several other probable collapse craters have formed on the outer boundary of the Rümker mass.

Marius Hills

McCAULEY (1967a, 1968) described the Marius Hills area (Fig.7–9) as a southward dipping plateau (area approximately 35,000 km^2) standing several hundred metres above the surrounding mare plain. The area takes its name from the large, mare-filled

Fig.7. Oblique Lunar Orbiter 2 photograph of part of the Marius Hills complex. (NASA.)

crater at the eastern edge of the complex. The surface is characterised by numerous, closely spaced domes ranging from 3 to 10 km across. They stand on the smooth to gently undulating surface of the plateau. Although the albedo of the Marius Hills unit—identified by McCauley as a stratigraphic unit and termed by him the Marius Group— is similar to that of the surrounding mare plain, it has a lower crater density which suggests it is younger; but it is overlain by rays from Kepler showing it to be older than Copernican. On this basis McCauley assigned to it an Eratosthenian age.

Different types of volcanic feature occur in the region. This suggests that lavas of different composition were erupted there. Most of the features resemble terrestrial volcanic forms in both shape and size. There is no necessity to propose eruptive processes different from those known on Earth to explain the majority of

the features. Superimposed on the features interpreted as being igneous in origin are what are presumably primary impact craters which tend to mask the endogenous craters. There are also clusters of craters associated with rays and these craters may be of secondary impact origin. Fig.9 is a geologic sketch map of part of the Marius Hills area prepared by the author from one medium-resolution Orbiter 5 frame (Fig.8); only features thought to be of igneous origin are shown.

On the scale of the medium-resolution photograph (Fig.8) domes may be classified as being of two types: low domes and steep sided domes. The low domes have heights of 50–100 m (KARLSTROM et al., 1968). Their surfaces are smooth and the outer boundaries are defined by the break in slope at their base. The slopes of the low domes are convex upwards and the domes tend to be circular in plan. In one or two cases scarps similar to flow fronts may be seen on the

Fig.8. Part of the Marius Hills complex, photographed by Lunar Orbiter 5. (NASA.)

lower flanks of these domes but, in general, their surfaces appear featureless on the photographs. Summit
pits are not present, although collapse pits are scattered
across a number of the domes. The lack of visible flows
may support the view of some authors that these
domes were formed by arching of the surface above
high-level laccolithic intrusions. However, bearing in
mind the known low viscosity of some lunar lavas, it
does not seem improbable that the low domes are
local extrusive centres analogous to terrestrial shield
volcanoes.

The steep sided domes may be up to 250 m high
(KARLSTROM et al., 1968). They are bounded by well
defined, convex upward scarps, whilst the tops tend
to be level or gently convex. Differing from most of
the low domes the steep ones have less regular bound-

aries and generally rough top surfaces. Many of the
steep sided domes are characterised by ridged surfaces
arranged, commonly, in a pattern resembling flow
ridging developed on thick, viscous terrestrial lavas
(see *1* in Fig.9). Others have low rimmed craters interpreted as having been formed by collapse while the
magma was still partially fluid. Domes of this type
may be interpreted as extrusions of viscous lava, either
as volcanic domes, or lava flows, or both. They have
volumes of up to a few cubic kilometres.

Often associated with steep sided domes are steep
sided cones that resemble the pyroclastic cones found
in many volcanic fields of the Earth. The lunar cones
may be single (see *3* in Fig.9) with one summit crater,
or multiple and aligned with one of the regional
structural directions (*2* in Fig.9). Cones on the smallest,

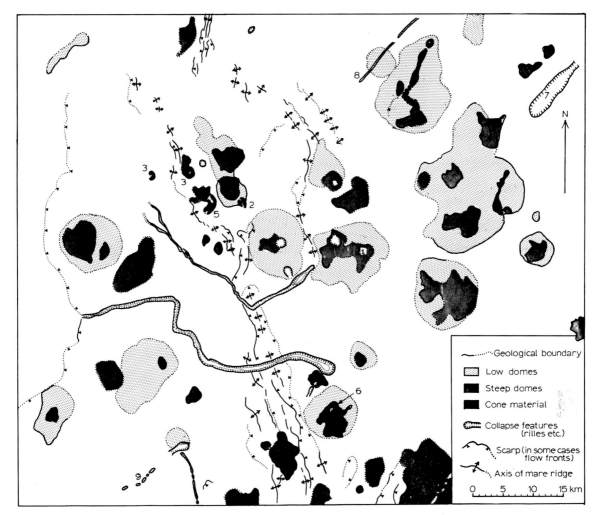

Fig.9. Geologic sketch map of the Marius Hills area (Fig.8). Only features thought to be volcanic are shown.

steep sided domes (*4, 5* and *6* in Fig.9) appear to mark the source vents for the dome material. Number *5* (Fig.9) is a breached cone (*C* in Fig.10) from which a double-lobed, steep sided flow *F* emerges. Similar relations may be observed at *4* and *6* (Fig.9); number *6* is the larger, and the flow has a ridged surface and a lobate front.

Rimless depressions thought to be of volcano-tectonic origin occur in several parts of the Marius Hills area. One of the largest is the feature numbered *7* in Fig.9. This is a trough-like feature with almost parallel walls in its longer direction; these walls follow one of the regional trends. To the west of this, at *8* in Fig.9, is a narrow rille, which also appears to have formed by collapse trending in the same direction. There is a short break in the rille near its northern end. Rilles (for example, *9* in Fig.9) in the southwest corner of the map are markedly discontinuous and consist of closely spaced, elliptical craters resembling a string of sausages. Such features have been discussed and com-

pared with terrestrial features, by J. B. Murray (Chapter 3): they are probably de-roofed underground channels in the lavas.

Mare ridges are well developed in the area, forming a complex running in a northerly to northwesterly direction. Two types of ridge may be defined (KARLSTROM et al., 1968): (*a*) broad ridges with wide flanks often terminating in a low scarp parallel to the ridge direction; and (*b*) narrow, steep ridges forming en echelon or braided patterns, commonly situated on broad ridges.

The main, broad mare ridge running northward from the bottom centre of the map (Fig.9) is cut by a large sinuous rille (see Chapter 3). MCCAULEY (1968) has pointed out that the elliptical cratered source of the rille to the north is partly surrounded by a raised rim, implying that it may be a volcanic vent. Again, there is some evidence of a flow front (*FF* in Fig.8; see also Chapter 3, fig.1) perpendicular to the rille at its lower, western end; this supports the view that the

Fig.10. High-resolution Lunar Orbiter 5 photograph of a single flow in the Marius Hills complex. Note the steep cone C over the vent, and the lobate front of the flow F. Arrows show the directions of flow. (NASA.)

rille is a collapsed lava channel. If this origin for the rille were accepted then some constraints would be placed on suggestions for the origin of the broad ridge that it cuts. As the lava in which the rille was formed would not have flowed uphill over the ridge, the formation of the ridge must post-date the lava. It is thus unlikely that the ridge is an extrusive phenomenon as the extrusive material would have invaded the rille, which is clearly not the case. It therefore appears that the ridge formed by an uparching of the surface, presumably above a subsurface dyke intrusion. As further evidence of this the floor of the rille also appears to be arched over the ridge. The narrow ridges are more likely to be extrusive, and some of these possibly extend into the rilles.

Conclusion

From the preceding descriptions it can be seen that a variety of features of igneous origin have formed in the lunar maria; in areas such as the Marius Hills several igneous features have developed in close association with one another.

Different types of eruption have therefore occurred depending on the physico-chemical conditions of the magma in the active vent. Variations in temperature, viscosity, gas content, rate of emission and composition will give different types of eruption and, therefore, different volcanic edifices. One of the principal factors governing the type of volcanic end-product will be the composition of the magma being erupted.

References

CARR, M. H., 1969. *U.S., Geol. Surv., Map I-599.*

FIELDER, G., 1961. *Structure of the Moon's Surface.* Pergamon, London, Chapter 12.

FIELDER, G., 1965. *Lunar Geology.* Lutterworth, London.

FIELDER, G. and KIANG, T., 1962. *Observatory,* 82: 8.

GUEST, J. E. and FIELDER, G., 1968. *Planetary Space Sci.,* 16: 665.

KARLSTROM, T. N. V., MCCAULEY, J. F. and SWANN, G. A., 1968. *U.S., Geol. Surv., Interagency Rept., Astrogeology,* 5.

MCCAULEY, J. F., 1967a. In: S. K. RUNCORN (Editor), *Mantles of the Earth and Terrestrial Planets.* Wiley, London.

MCCAULEY, J. F., 1967b. *U.S., Geol. Surv., Map I-491.*

MCCAULEY, J. F., 1968. *Am. Inst. Aeronaut. Astronaut.,* 6: 1991.

MURASE, T. and MCBIRNEY, A. R., 1970. *Science,* 167: 1491.

RIFAAT, A. S., 1967. *Icarus,* 7: 267.

SALISBURY, J. W., 1961. *Appl. J.,* 134: 126.

SALISBURY, J. W., ADLER, J. E. M. and SMALLEY, V. G., 1967. *Dark-haloed Craters on the Moon.* U.S.A.F. Cambridge Research Laboratories Preprint, Bedford, Mass.

WILHELMS, D. E., 1968. *U.S., Geol. Surv., Map I-548.*

5

Multiphase eruptions associated with the craters Tycho and Aristarchus

R. G. STROM AND G. FIELDER

Introduction

Tycho, the well known lunar crater 85 km in diameter, and Aristarchus, the brightest lunar crater and 40 km in diameter, are two of the most prominent craters on the Moon. Both craters have clearly sculptured rims, are relatively deep, and are centers of stratigraphically recent rays. The craters are widely regarded as recent impact structures. In addition, Tycho presents an anomalously high radar reflectivity (PETTENGILL and THOMPSON, 1968) which indicates that the crater is rougher or denser than the surroundings; and, during eclipse or through a lunation, Tycho also follows anomalous thermal curves (SAARI and SHORTHILL, 1965) which indicate that the materials of the crater are better thermal conductors than the surrounding terrain. All these observations may be explained on the impact hypothesis.

In 1967 part of Tycho and all of Aristarchus were photographed from Orbiter 5 with a ground resolution of about 5 m. In addition, Orbiter 5 secured medium-resolution photographs of the whole of both structures and their immediate surroundings, with a ground resolution of about 40 m. It was therefore possible to study some parts of the Tycho and Aristarchus domains in considerable detail and to link these studies through a general description of the surrounding terrain, in each case.

Geologic units of Tycho

The region north of Tycho consists of various deposits with distinct surface morphologies. Fig.1 is a generalised map of the different geologic units, and Fig.2 is a mosaic of three medium-resolution photographs showing the area mapped in Fig.1.

Units Tf_1 and Tf_2 are relatively thick, massive bodies with a ropy or ridged surface containing numerous open fractures and abundant blocks (Fig.3, 4). Judging by the shadows cast by the fronts of the units Tf_1 and Tf_2, they are of the order of 40 m and 20 m thick, respectively. The margins have raised rims similar to the natural levees associated with lava drainage channels. In fact, the eastern edge of unit Tf_2 forms a well defined front which has overflowed the channel rim and spilled over the adjacent terrain. Since the height of the rim (~ 60 m) is considerably greater than the thickness of the flow front, it appears that Tf_2 is the latest of several flows or surges which have issued from the channel and built the natural levee.

Both units Tf_1 and Tf_2 have higher albedoes than the adjacent area, as shown in Fig.5. Unit Tf_3 is similar to units Tf_1 and Tf_2 but is thinner (~ 12 m), has a lower albedo, and is somewhat less ridged (Fig.3). The ridging on these units is indicative of pressure ridging formed by the flow of a viscous fluid, and is similar to the surface of viscous lava flows (Fig.6, 7) on Earth. The units Tf_1, Tf_2 and Tf_3 overlie Ta and Tb, and are therefore younger than them. Also Tf_1 overlies Tf_3 and is therefore younger than that unit. The surface texture, the well defined flow fronts and the flow channels strongly suggest that all these units are lava flows which were relatively viscous. Units Tf_1 and Tf_2 appear to have been more viscous than other units in the area. Also, Tf_1 and Tf_2 have a higher albedo than that of the neighbouring units— and higher, in particular, than the albedo of the unit on which Sur-

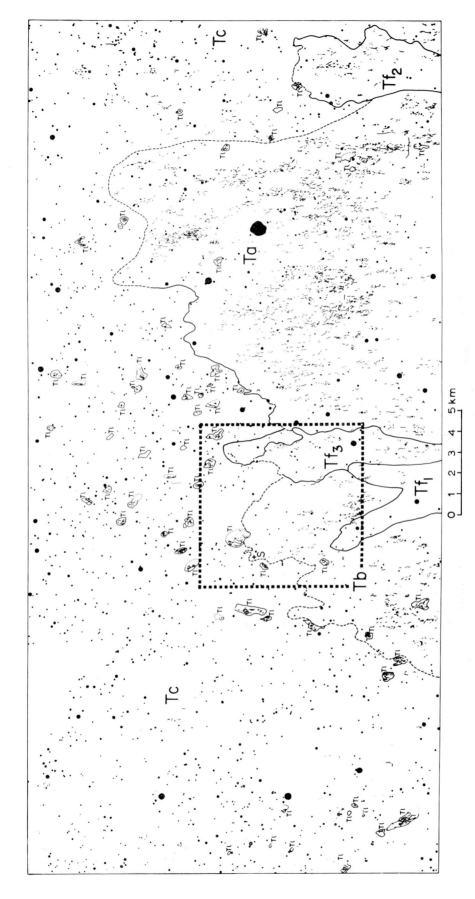

Fig.1. Generalized map of the geologic units north of Tycho showing the distribution of craters (filled circles), fractures (fine lines) and "lakes" *(Tl)*. The dashed outline indicates the location of the detailed geologic map (Fig.13). *S* marks the location of Surveyor 7. (Compiled from Orbiter 5 photo.)

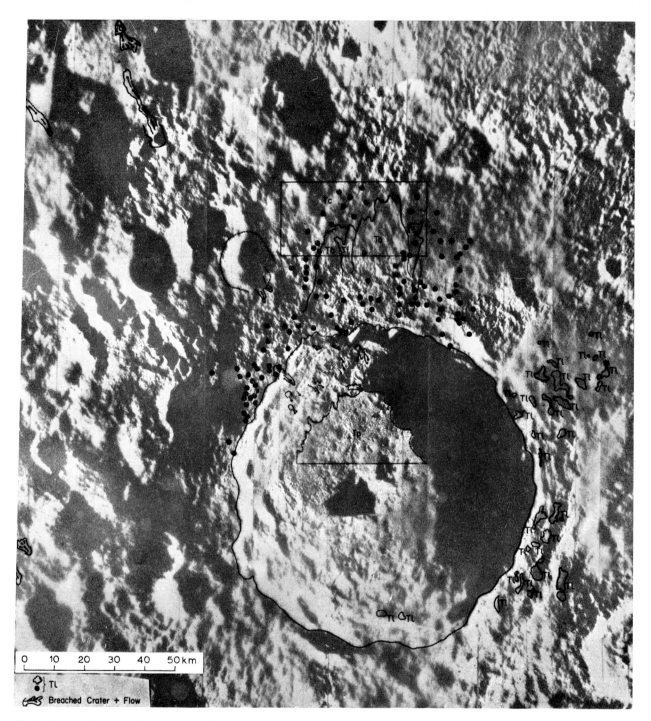

Fig.2. Mosaic of three Orbiter 5 photographs. The rectangle north of Tycho outlines the area mapped in Fig.1, and that portion of the floor of Tycho mapped in Fig.23 is also shown. The black dots give the approximate location of those "lakes" on which craters were counted. Larger "lakes" are shown in outline. The principal flows outside the mapped areas are sketched, and the directions of flow indicated by arrows. (NASA.)

veyor 7 rests—so Tf_1 and Tf_2 may well be more acidic than the anorthosite indicated by the Surveyor 7 analysis.

Units Ta and Tb are characterised by a high density of fine fractures shown as fine, short lines in Fig.1, and by a semi-blocky, slightly ridged appearance. The units are generally smoother than units Tf_1, Tf_2 and Tf_3. In the southern portion of the mapped area the texture is similar to that of units Tf_1, Tf_2 and Tf_3, but somewhat more subdued. Both Ta and Tb overlie and are topographically higher and less cratered than unit Tc and, therefore, must be younger than that unit.

Fig.3. Flow units Tf₁ and Tf₃. Notice the ropy or ridged texture and well defined fronts of these units, and the open channel on unit Tf₁. (Orbiter 5 photo, NASA.)

The fracturing on Ta and Tb is primarily associated with the crests of ridges and, although fracturing occurs in the troughs between ridges, it is much less intense than the fracturing of the Tf units. Thus the fracturing of Ta and Tb may be due to uplift of the ridges possibly resulting from the hydrostatic pressure of fluid lava beneath a solidified crust. All the above characteristics suggest that units Ta and Tb are volcanic flows. They are probably composed of numerous flow units.

Unit Tc is a relatively smooth, hummocky material largely devoid of fractures, and more densely cratered than the other units. All other mapped units overlie it and, therefore, it is the oldest unit in the area. Two sets of closely spaced ridges occur in Tc. The more prominent set is roughly concentric with Tycho whilst the other, weaker, set is approximately orthogonal to the first set. This type of ridging is absent or developed only weakly and locally on the other units.

The set of ridges that is sub-radially oriented with respect to Tycho probably owes its origin to fracturing. Several large, sharp craters on unit Tc, as seen from Surveyor 7, are not surrounded by blocks, whereas other smaller, sharp craters in the vicinity of the large craters have very blocky rims. This indicates that unit Tc consists of more or less randomly distributed areas of fine, weakly cohesive material and hard, dense, coarse or massive material; whilst unit Tb consists entirely of dense material overlain only by a thin layer of fragmental debris.

In addition to the above units, there are numerous "lakes" consisting of relatively dark, smooth material

Fig.4. Flow unit Tf₂. The flow has overriden the channel rim *r*—probably a natural levee—and spilled onto the adjacent terrain to the east forming a well defined front *(f)*. (Orbiter 5 photo, NASA.)

exhibiting fine fractures. They are situated in circular to irregular depressions and are concentrated within 20 km of the rim of Tycho. The fracture pattern on many of the lakes (Fig.8) is remarkably similar to that found on lava lakes (Fig.9) in Hawaii, where the pattern results from tensile stresses caused by thermal contraction during cooling of the lava. The polygons formed by the fracture patterns are conspicuously convex upward in both the lunar and terrestrial lakes. According to PECK and MINAKAMI (1968) the terrestrial polygonal hummocks result from the accumulation of gases trapped beneath the crust near the centers of the polygons, and the escape of gases from marginal parts. This causes upbowing of polygon centres and downsagging of margins. The similarity in pattern and shape of the lunar and terrestrial polygons strongly indicates that the lakes in the vicinity of Tycho have cooled from originally molten material.

Several of the larger lakes have distinct fronts where they contact the sides of the depressions in which they lie and, in a few cases, the lakes have spilled over low portions of their confining basins forming short, elevated flows (Fig.10,11) in the adjacent terrain. At least one lake has apparently experienced two periods of eruption as evidenced by a secondary flow superposed on the lake surface and indicated by the arrow in Fig.11. The secondary flow appears to have issued from vents or fissures at the northeastern edge of the lake. All these facts strongly suggest that the lakes consist of lava which has found egress to the surface through vents or fractures primarily associated with the depressions in which they lie. The distribution of the lakes near the rim of Tycho indicates that they are intimately associated with its formation.

A small lake (Fig.12) 350 m in diameter occurs in an irregular crater near the crest of the main central peak of Tycho. The morphology of the lake is similar to that of the other lakes and it even displays fractures or crater chains on its surface. It seems likely that the lake formed in the same manner as the others and, if this is so, then it is probable that the crater containing the lake is a volcanic vent connected by a conduit with

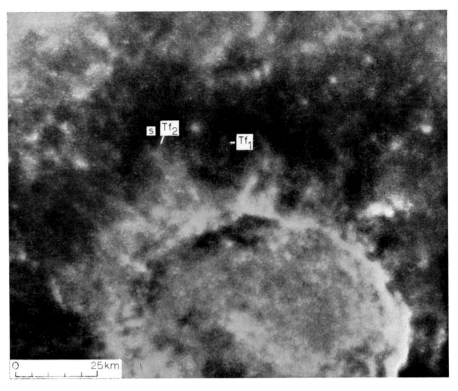

Fig.5. Full-Moon photograph of the northern part of Tycho and vicinity showing the high albedo units Tf₁ and Tf₂, and the approximate location of Surveyor 7 *(S)*. (Catalina Observatory photo, Consolidated Lunar Atlas.)

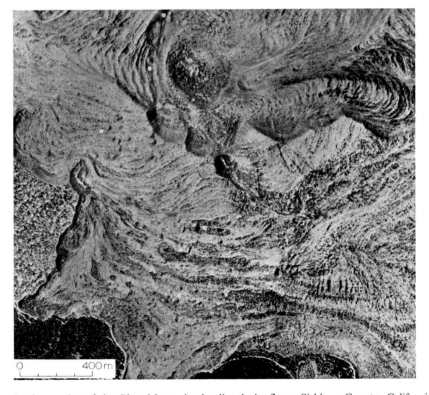

Fig.6. Aerial photograph of a portion of the Glass Mountain rhyolite–dacite flows, Siskiyou County, California, showing numerous explosion pits. Several of the larger explosion craters are indicated by "*e*". (U.S. Department of Agriculture photo.)

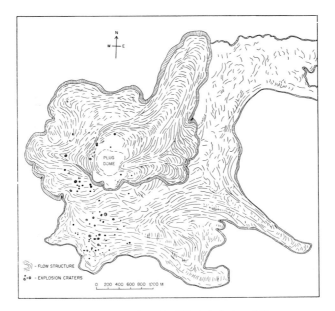

Fig.7. Map of the Glass Mountain flows (Fig.6) explosion pits (filled circles) and flow structure. (Mapped by R. G. Strom.)

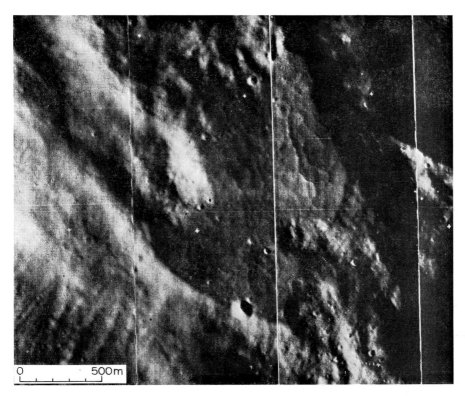

Fig.8. "Lake" near Tycho showing fracture pattern. (Orbiter 5 photo, NASA.)

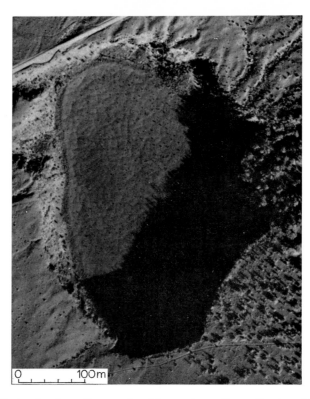

Fig.9. Keanakakoi crater (Hawaii), showing the pattern produced by cracks on the cooling lava lake. The floor is covered by about 15 cm of pyroclastics from the 1959 eruption of Kilauea Iki. (U.S. Department of Agriculture photo.)

a magma source. In this event it is possible that the central peak itself is of volcanic origin.

Since the lakes have a paucity of craters and interrupt the other units, they must be relatively recent features of post-Tycho age. Crater counts reported in this chapter indicate that the average age of the lakes is close to that of the floor of Tycho. However, all lakes must be younger than unit Tc because the lakes interrupt this unit which is the oldest unit in the mapped area.

In order to clarify the stratigraphic relations in the well defined, morphologically different units in the vicinity of the Surveyor 7 spacecraft, a map (Fig.13) of the area was prepared. The map shows the various geologic units in detail, and the distribution of rocks, craters and fractures. Units A and B are the same as units Tf_1 and Tf_3 and have been discussed previously. Units C, D, and E are the same as Tb. Detailed mapping of this area has shown that unit C has a unique morphology, overlies unit E, and therefore should be separated from Tb. The stratigraphic position of these units is well defined and shows that: (1) A is the youngest unit and overlies B, C and E; (2) B is older, overlying D and C; (3) C is still older and overlies E; (4) E overlies F; and (5) F is the oldest unit. The stratigraphic position of D relative to units C and E

is not clear, but the surface morphology suggests that it may be related to, and of the same age as, unit E.

The relation between unit C and the largest lake in the area bears on the age and origin of units A, B and C. Fig.14 shows that the northern terminus of unit C overlaps and extends about 400 m over the southern portion of the lake. There appear to be several collapse depressions, d, near the end of the unit where it covers the lake. Since the lake interrupts unit F and must be younger than that unit, then unit C, which overlaps the lake, must be younger than unit F; and units A and B, which overlap unit C, must be still younger. As previously indicated, all lakes are post-Tycho in age and, therefore, units A, B and C must also be of post-Tycho age. This indicates that these units, at least, are volcanic flows which were erupted after the formation of Tycho.

Table I lists the number density of blocks over about 4 m in diameter on the various units in Fig.13 and the number of craters per square kilometre greater than 20, 40, 60 and 80 m in diameter, respectively. There appears to be a negative correlation between the number of blocks on a given unit and the density of craters on the unit. This indicates that the majority of blocks greater than about 4 m in diameter are not ejecta from the craters on the various units.

Fig.10. "Lakes" on the northeastern rim of Tycho. Note the well defined fronts *(f)* in two of the lakes. The material of two "lakes" *(A* and *B)* has apparently originated at a higher level and cascaded down-slope to form flow fronts on lower terrain (arrow). One of these flows overlaps a pre-existing lake *(B)*. (Orbiter 5 photo, NASA.)

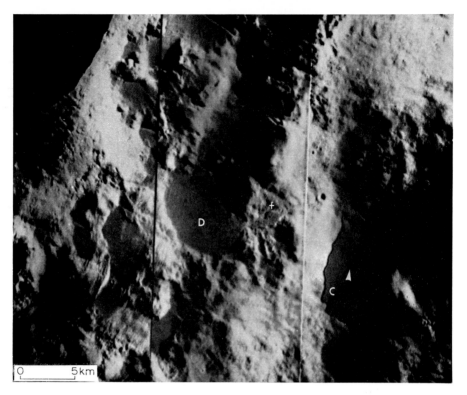

Fig.11. "Lakes" on the southeastern rim of Tycho. Note the well defined fronts and the secondary flow (arrow) superposed on the surface of lake C. A short flow with a steep front (f) has issued from lake D through a low section of the confining basin. (Orbiter 5 photo, NASA.)

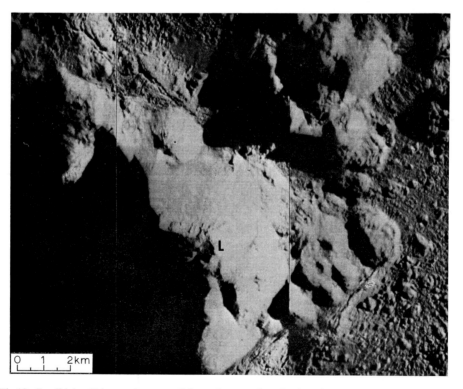

Fig.12. Small lake (L) near the crest of the main central peak of Tycho. (Orbiter 5 photo, NASA.)

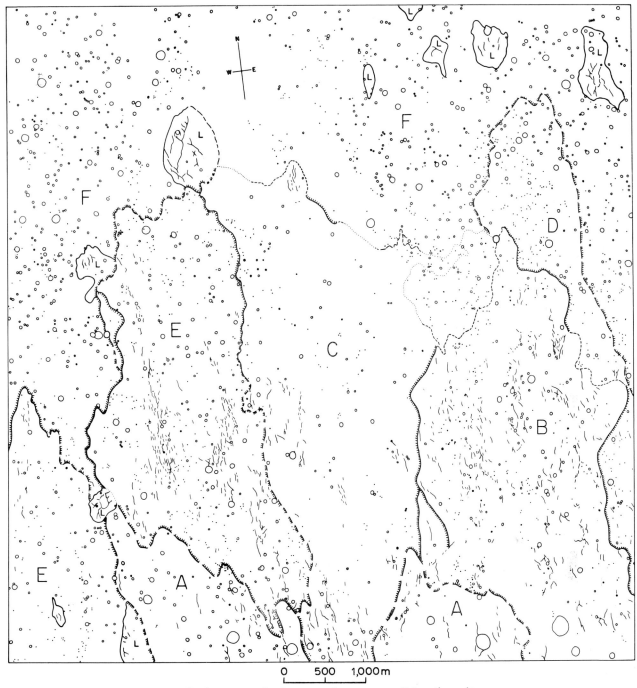

Preliminary geologic map of surveyor 7 landing site

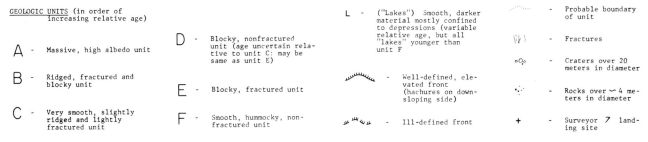

GEOLOGIC UNITS (in order of increasing relative age)

A - Massive, high albedo unit

B - Ridged, fractured and blocky unit

C - Very smooth, slightly ridged and lightly fractured unit

D - Blocky, nonfractured unit (age uncertain relative to unit C: may be same as unit E)

E - Blocky, fractured unit

F - Smooth, hummocky, non-fractured unit

L - ("Lakes") Smooth, darker material mostly confined to depressions (variable relative age, but all "lakes" younger than unit F

- Probable boundary of unit

- Fractures

- Craters over 20 meters in diameter

- Rocks over ~ 4 meters in diameter

- Well-defined, elevated front (hachures on down-sloping side)

- Ill-defined front

+ - Surveyor 7 landing site

Fig.13. Detailed geologic map of the area located by the dashed outline in Fig.1, in the immediate vicinity of Surveyor 7.

Fig.14. Orbiter 5 photograph showing the terminus of unit *C* overlapping a lake. Note the well defined front and possible collapse depressions *(d)* near the distal end of the flow. (NASA.)

TABLE I

BLOCK DENSITY COMPARED WITH CRATER DENSITY ON THE GEOLOGIC UNITS
SHOWN IN FIG.13

| Unit | Craters per km² | | | | Blocks per km² |
	$\geqslant 20\ m$	$\geqslant 40\ m$	$\geqslant 60\ m$	$\geqslant 80\ m$	$\geqslant 4\ m$
B	11.89	2.01	0.51	0.38	34
C	10.02	1.45	0.36	—	19
D	22.54	4.64	2.50	0.71	67
E	21.32	3.10	0.80	0.10	39
F	33.32	4.32	1.12	0.42	19

The blocks are probably a combination of: (*1*) ejecta from Tycho; and (*2*) blocks of lava broken up and carried by the flow process. The latter possibility may explain the tendency to non-random distribution of blocks with respect to the various units. A rock slab (Fig.15) approximately 50 cm long and 15 cm thick, photographed by Surveyor 7 on unit *E*, may

this area; and the flows are associated with the breached craters. In addition, there are several strings of vents from which flows have emanated. These craters, and possibly concealed fissures, are probably the sources of the flows which have breached their walls and flowed in a northerly direction.

Several additional flow structures occur outside

Fig.15. Possible slab of lava photographed from Surveyor 7 on flow unit *E*. Note rows of vesicles on edge of slab. (NASA.)

be one such block of lava. The block has a smooth, slightly concave underside and an irregular top. Several rows of vesicles up to 1 cm in diameter parallel the long axis of the slab.

The sources of the flows described appear to be in the region (Fig.16) just north of the rim of Tycho. This is an anomalous area where several irregular craters, 0.5–2 km in diameter, have their northern rims breached. Extensive flows with well defined fronts are found in

the mapped areas and are shown in Fig.2 and Fig.17. They are 3–20 km long and about 1–6 km wide. The direction of flow is indicated by the arrows in Fig.2 and conforms to the slope of the terrain. Several flows ran towards Tycho. Others flowed away from Tycho. Five of these six flows seem to have issued from craters which have been breached by the flows. Their surface texture is coarse, with a wrinkled or ropy appearance, and several seem to have been considerably more

Fig.16. Probable source area of the flows mapped in Fig.1 and Fig.13. The flows occur just off the top of the photograph. Several possible volcanic vents with their northern rims breached are indicated by V, and flows associated with these craters are designated f. The rim of Tycho is just off the bottom of the photograph. (Orbiter 5 photo, NASA.)

Fig.17. Thick, viscous flow which has issued from a large crater *(C)* mostly off photograph. This flow has travelled towards Tycho. (Orbiter 5 photo, NASA.)

Fig.18. Flows on the southern interior wall of Tycho. One flow (f_1) has originated at a lake on the wall (off the bottom of the photograph) flowed down the wall (arrows) and bifurcated at *A*, one half flowing onto the floor at *B* and the other half issuing onto a level portion of the wall at *C*. Other flows are indicated by f_2 and f_3. (Orbiter 5 photo, NASA.)

Fig.19. Northern interior wall of Tycho showing flows and flow channels *(f)* several of which issue from breached craters *(c)*. (Orbiter 5 photo, NASA.)

viscous than the flows in the maria.

Fig.17 shows a particularly viscous-looking flow with a high front (~ 80 m), and with arcuate and linear flow structure. This flow has apparently issued from the crater (*C*), partially shown in the upper left of the photograph, and has flowed towards Tycho. The general appearance of the flows and the fact that they are associated with craters indicate that they are volcanic lava flows. The most distant flow is found near to the limit of the medium-resolution coverage of Orbiter 5 in the area, just over 100 km from the rim of Tycho. It is possible that other flows occur at even greater distances from Tycho.

The interior slopes of Tycho also show signs of the flow of material (Fig.18–20). These flows are primarily expressed as channels with raised rims, although there are several flows with well defined fronts. Several flows (Fig.18) have originated at lakes and craters on the wall, flowed down valleys in the wall and ponded in low areas of the wall or issued upon the floor. The flows display arcuate ridging and other flow structures similar to those found on terrestrial lava flows. Since several of the flows have originated at lakes, and the volumes of the flows are large, the lakes are surely volcanic centres which have experienced relatively sustained eruptions.

Fig.21 shows a flow which has issued from a pond at a higher level, flowed across part of the floor and merged with it so that the floor and flow appear as one textural unit. Peripheral fissures on the floor continue across the flow. Thus the flow reached the floor before the formation of the peripheral fissures, and the flow must be essentially the same age as the floor.

The flow channels have raised rims and several appear to begin at breached craters (Fig.19,20) on the wall. The channels appear to be identical to lava drainage channels with natural levees built up by the overflow of lava. The similarity is strengthened by the fact that at least one of the channels ends in a small "lake" which exhibits a flow front.

Finally, one part of the floor (Fig.22, 23) of Tycho forms a geologic unit which is morphologically quite different from most of the other units. It is characterised by steep and gentle sided domes (not mapped) interspersed with terrain which displays a highly fractured and ropy pattern similar to that of terrestrial lava. However, the sizes of the fractures and ridges are at least two orders of magnitude greater than similar features on terrestrial lava deposits. This is probably the result of the lower lunar gravity which would allow larger fractures and pressure ridges to develop. The

Fig.20. Flows *(f)* on the northern interior wall of Tycho. One of the flows has apparently originated in the smooth material outside the rim and flowed down the rim scarp (arrows). The breached crater *(c)* on the rim seems to have given rise to much of the smooth material (lava) just outside the rim. (Orbiter 5 photo, NASA.)

Fig.21. Part of the ropy material *(r)* of a pond has issued onto the floor of Tycho as a short flow *(f)* which has been fractured by subsequent subsidence on the periphery of the floor. (Orbiter 5 photo, NASA.)

Fig.22. A portion of the northern floor of Tycho showing ropy and fractured texture, possible tumuli *(T)* and probable small volcanic vents *(V)*. (Orbiter 5 photo, NASA.)

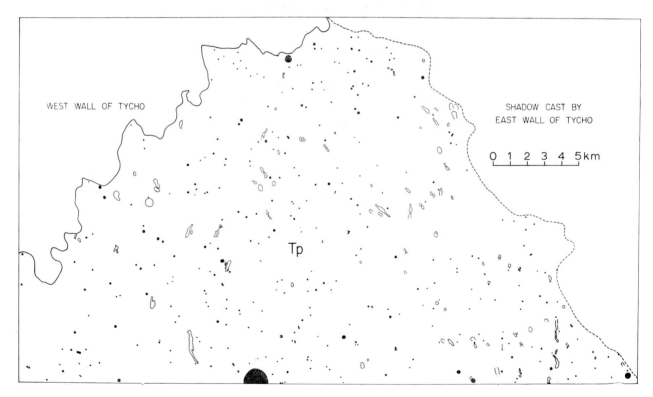

Fig.23. Map of a portion of the floor of Tycho compiled from Orbiter 5 photograph. The filled circles represent eumorphic craters of probable impact origin. Probable volcanic vents are shown in outline. (NASA.)

texture and albedo of the floor is similar to that of units Tf_1 and Tf_2 and, therefore, they may have a similar composition; that is, more acidic than other flow units. Several gentle domes exhibit the same texture as the floor and have fissures along their crests. They are probably updomings of the floor or tumuli produced by the hydrostatic pressure of fluid lava beneath the indurated crust.

Numerous crater chains which trend in three dominant directions, and craters of highly irregular shape, occur on the floor. Several of these features are shown in Fig.22 and 23 and attest to the intense volcanic action to which the floor has been subjected.

A set of peripheral fractures found on an inward sloping bench at the margins of the floor and central peaks, and floor material occurring at higher levels on the western wall, indicate that the floor was once at a higher level than at the present time and has subsequently subsided.

Chemical analyses from the Surveyor craft

Surveyor 7 landed on the extreme edge of flow unit E (Fig.1,5,13), and chemically analysed the fragmental material and one rock by an alpha scattering device (PATTERSON et al., 1970). The results of these analyses, and those from Surveyors 5 and 6 in Mare Tranquillitatis and Sinus Medii, respectively, are presented in Table II together with the analyses of the Apollo 11 and 12 rocks.

The chemical composition of rocks at the Surveyor 7 site is quite different from those at the Surveyor sites in the maria and from the analyses of the basaltic rocks removed from the Apollo 11 and 12 sites in

Mare Tranquillitatis and Oceanus Procellarum, respectively. However, the composition of rocks analysed through Surveyor 7 is close to that of the small fragments of anorthosite found in the soil at the Apollo 11 site by WOOD et al. (1970). This anorthosite is thought to have derived from the highlands. It follows that flow unit E may well have an anorthositic composition. The densities of the lunar anorthosites are between 2.8 and 2.9 g/cm^3, whilst the densities of the lunar basalts are about 3.4–3.5 g/cm^3. Anorthosites are leucocratic rocks and it is probable that this accounts for the higher albedo of the highlands compared with that of the maria. If an anorthositic composition is characteristic of the highlands then widespread chemical differentiation and/or crystal fractionation have taken place on the Moon. The Apollo 12 sample 12013 (Table II) is a siliceous rock with an exceptionally high potassium content and has a composition similar to an andesite or basaltic andesite. This sample again indicates that rather extensive magmatic differentiation has occurred on the Moon.

Again, it has been reasoned that the volcanic flows Tf_1, Tf_2, and the flow shown in Fig.17, are more acidic than the flow unit E. Therefore, extensive magmatic differentiation may have been involved in the later stages of the development of Tycho.

Crater counts in and near to Tycho

If it is supposed that each geologic unit on the Moon is a counter of primary impact craters, then it is clear that older units will be more densely cratered than recent ones. In fact, the craters may belong to three different populations—primary impact craters,

TABLE II

CHEMICAL COMPOSITION AT THE SURVEYOR 5, 6 AND 7 SITES AND OF THE APOLLO 11 AND 12 ROCKS (IN WEIGHT %)

Oxide	Surveyor 5	Apollo 11 basalts (average)	Surveyor 6	Apollo 12 basalts (average)	Surveyor 7*	Apollo 11 anorthosites	Apollo 12 sample 12013
SiO_2	46.4	40.8	49.1	40.0	46.1	45.7	61.0
TiO_2	7.6	10.5	3.5	3.7	—	0.3	1.2
Al_2O_3	14.4	10.0	14.7	11.2	22.3	30.5	12.0
FeO	12.1	18.8	12.4	21.3	5.5	4.5	10.0
MgO	4.4	7.5	6.6	11.7	7.0	4.8	6.0
CaO	14.6	10.9	12.9	10.7	18.3	15.8	6.3
Na_2O	0.6	0.48	0.8	0.45	0.7	0.35	0.69
K_2O		0.21		0.06		—	2.0
MnO		0.22		0.26		0.1	0.12
H_2O		0.005					

* The rock analysed at the Surveyor 7 site had essentially the same chemical composition as the undisturbed soil except for a lower magnesium and higher aluminium content, and a slightly lower iron content.

secondary impact craters (from either a meteoric impact or a volcanic eruption), and craters of internal origin. For the present, the results of counting all eumorphic circular craters as large as, or in excess of, 50 m in diameter (Fig.23) will be presented.

The counts were performed for each geologic unit separately and the total area of each unit was measured using a planimeter. Parts of all photographed areas were overexposed or shadowed, so that the actual areas on which craters were counted were less than the total areas measured. Corrections to the total areas were applied using planimetric and square-counting methods.

Although the counts of craters on the floor of Tycho are restricted to eumorphic craters, the position of many at the apices of hills and on fissures suggest that they may be volcanic rather than impact in origin: therefore, the count on this unit (Tp) represents a maximum number of craters. The measured areas and counts for the different units are given in Table III.

TABLE III

TOTAL CRATER COUNTS
IN DIFFERENT GEOLOGIC UNITS OF TYCHO

Geologic unit	Areas examined (km²)	Corrected area of count (km²)	Total number of craters ⩾ 50 m in diameter	Craters per unit area (floor = 1)
Lakes	22	22	17	0.9
Floor	460	400	347	1.0
Tf₁ + Tf₃	24	19	30	1.8
Tf₂	19	15	22	1.7
Ta	185	136	278	2.4
Tb	51	42	81	2.2
Tc	463	342	1,324	4.5

The geologic units Tf_1, Tf_2 and Tf_3 are alike and, indeed, are found to have the same crater density within the limits to be expected from the method. Again, the apparently similar units Ta and Tb have similar number densities of craters. A less obvious result of the crater counts is that the "lakes" (Tl), indicated by filled circles in Fig.2, and the floor (Tp) have a similar number density of craters (Table III). Unit Tc has 4.5 times as many craters per unit area as the floor of Tycho, 2.6 times as many as units ($Tf_1 + Tf_2 + Tf_3$) and twice as many as unit (Ta + Tb).

Consistently, more craters per unit area are found on stratigraphically older units. It would appear probable, therefore, that the majority of the eumorphic craters are impact craters.

The data are represented logarithmically (Fig.24) as cumulative numbers per 1,000 km² of surface against

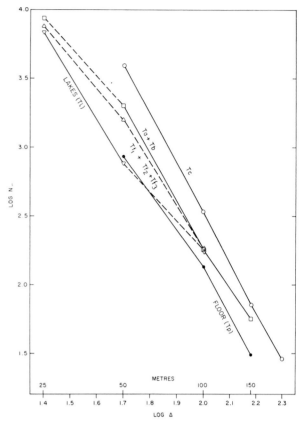

Fig.24. Logarithmic plot of the cumulative number (N) of craters per 1,000 km² of surface against minimum crater diameter (Δ) for the geologic units in Fig.1, 2 and 23.

minimum crater diameter in metres. The less significant points are joined with broken lines. Significance was judged qualitatively, using three criteria. The more significant points were defined as those satisfying three conditions simultaneously: the points were: (a) those relating to counts on areas generally larger than about 20 km²; (b) those in which the number of craters counted was > 10; and (c) those in which observational incompleteness was believed to be essentially absent. Point (c) was ascertained after maps of craters had been cross-checked by a second observer.

Because the ratio between the crater densities of units ($Tf_1 + Tf_2 + Tf_3$) and (Ta + Tb) was only 1.3 at the 50-m size (Table III), and was not based on the most significant data, we took sample counts in these regions of craters smaller than 50 m but larger than 25 m to see if the same difference persisted. The result (Fig.24) shows a similar difference in crater density, although we suspect some observational losses in the counts of the smallest craters since the terrains are rough and the graphs bend downward relative to the graph for the "lakes", where equally small craters were seen easily.

The fact that ($Tf_1 + Tf_2 + Tf_3$) is younger than

(Ta + Tb) derives from stratigraphic evidence. The counts are insufficient by themselves to establish the relative age of these particular units even if the craters counted are all impact craters; but, once again, the counts do not conflict with the geologic information. Again, therefore, we conclude that the majority of the eumorphic craters which we have used in this analysis are not of internal origin.

Counts of small craters near to Surveyor 7

Fig.25 is a logarithmic plot of the cumulative number of craters per 1,000 km² of surface against minimum crater diameter for the craters ≥ 20 m in diameter shown in Fig.13. The plot shows that, with the exception of units A and B, the number densities of craters on the units accord with the stratigraphic position of the units; that is, the older the unit the more craters there are per unit area. However, units A and B have an excess of craters in the diameter range 20–60 m. The youngest unit A has more craters per unit area than the oldest unit, F; and the next youngest unit, B, has more craters than unit C, which it overlaps. However, it was shown in the previous section that counts of craters over 50 m in diameter on the large areas shown in Fig.1 gave crater densities which were in agreement with observed stratigraphic relations. Since the areas of the units in Fig.13 are only 2–7 km², it is possible that the excess craters in this size range are due to random fluctuations in crater density. However, it is peculiar that an excess of craters is found only on the two most massive units and that the more viscous appearing of the two units (A) has the greater excess of craters. If the distribution of craters on these units is not due to random fluctuations, then the excess craters cannot be the result of primary or secondary impact and must be of internal origin.

In this case, it would seem likely that the excess craters were explosion pits resulting from the explosive escape of volatiles contained in the flows. It was pointed out previously that the flows in question have a surface morphology (flow ridging) similar to viscous flows on the Earth. Also unit A has a higher albedo than the other units. Terrestrial flows of acidic and intermediate composition often contain large amounts of volatiles which escape violently to form explosion pits. The Glass Mountain rhyolite–dacite flows (Fig.6) in Siskiyou County, California, have been mapped (Fig.7) to show the flow structure and distribution of explosion craters on the flows. If the excess craters on the lunar flows were of internal origin, then the slope of the

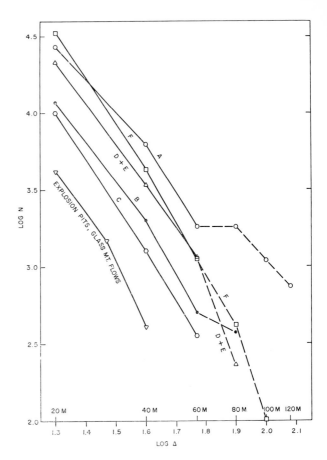

Fig.25. Logarithmic plot of the cumulative number (N) of craters per 1,000 km² of surface against minimum crater diameter (Δ) for the geologic units in Fig.13.

diameter–frequency curve might be similar to that of the terrestrial explosion pits. It is seen (Fig.25) that the slopes are indeed similar (about −3). The log N intercept of the Glass Mountain curve could be adjusted upwards, into the lunar domain, by the addition of impact craters. The fact that the excess of craters on unit A is greater than that on unit B may be due to unit A's having a higher volatile content. This possibility is substantiated by the more massive character and higher albedo of A relative to B.

Facts relevant to the interpretation of Tycho

In what follows, reference should be made to Fig.1–25.

(1) Counts (Table III) of craters ≥ 50 m in diameter on the various geologic units of Tycho demonstrate without exception that stratigraphically younger units contain fewer craters. Therefore it is probable that the majority of these craters are either primary or secondary impact craters. On two viscous appearing flows

there is an excess of small craters (< 50 m diameter). The excess craters may be internal explosion pits.

(2) The number of craters $\geqslant 50$ m in diameter per unit area of the floor unit (Tp) of Tycho is essentially the same as the average number per unit area on the lakes (Tl) outside the walls of Tycho. The floor and the lakes may therefore have the same average age.

(3) Major flows occur in the Tycho region. They are up to 20 km or more long and 4–12 km wide. Some appear to have been more viscous than others, and the former have a higher albedo than the others. Generally, the apparently more viscous flows are the later ones.

(4) The periphery of the floor of Tycho has an inward sloping bench containing fractures A similarly fractured bench surrounds the central peaks. In both cases the fractures concentrate in the zone of maximum curvature. They are probably tension fractures induced by the stretching of the margins of the floor, the lowest parts of which appear to have subsided.

(5) One flow, found on the inner wall of Tycho, is the same age as the floor.

(6) At least five flows have originated in, and breached, craters in the Tycho region. Several of these flows commence at distances of at least 100 km from the rim of Tycho. Other, less well defined, flows occur in the vicinity of the rim of Tycho. Like the major flows (3), some of the flows from craters near Tycho, and from craters in its rim, appear to have been more viscous than others. They are from 3 to 20 km long and up to 6 km wide.

(7) Of the "lakes", the major ones are concentrated in a relatively narrow zone on the eastern outer flanks of Tycho. Many of these lakes exhibit flow fronts.

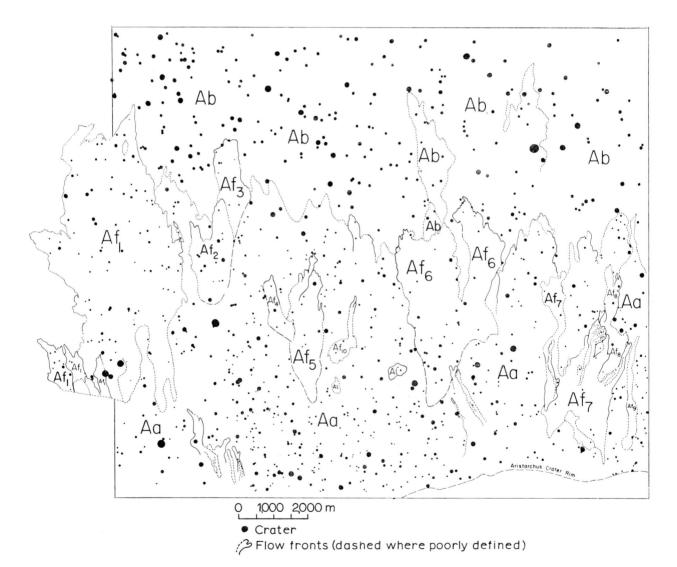

Fig.26. Generalized map of a portion of the north rim of Aristarchus showing the various geologic units and the distribution of craters greater than 28 m in diameter on units Aa and Af_1–Af_9, and greater than 56 m on unit Ab. (Compiled from Orbiter 5 photo.)

In at least two cases the lakes have overridden low places in the wall of their otherwise confining depressions and formed distinct flows. In several "lakes" there are younger flows with fairly well defined secondary fronts superimposed on the main flow. One lake occurs in the crest of the main central peak.

(8) Different units have marked differences in texture.

Geologic units of Aristarchus

The northern rim of Aristarchus contains flow units similar in disposition to those on the northern rim of Tycho. However, the surface morphology of these units is somewhat different from that near Tycho. The various geologic units of part of the northern rim of Aristarchus have been mapped in Fig.26. The map shows the distribution of craters greater than 28 m in diameter on units Aa and Af_1–Af_9, and greater than 56 m on unit Ab.

Units Af_1 and Af_7 are discrete, well defined flows which overlie, and are therefore younger than, units Aa and Ab. Af_1 (Fig.27) is an exceptionally smooth flow with very little fracturing or ridging. Af_7 has apparent leveed channels along its centre and at least one channel (Fig.28) has weak, arcuate ridging, indicative of flow ridging, at its terminus. All the other units have well defined, lobate fronts but are usually thin, ranging from less than 1 m to about 9 m in thickness. Lakes similar to those in the vicinity of Tycho appear to mark the source of several flows, and a row of vents (Fig.29, 30) is apparently the source of another flow. Other flows appear to have originated at small, irregular, flat-floored depressions, and still others have no evident source. Although the resolution of Earth-based full Moon photography is not enough to distinguish these flows, it is obvious that the general area in which they occur has a low albedo similar to that of the maria (see Fig.31). All the above characteristics suggest that the flows were extremely fluid.

Unit Aa consists of hilly, ridged terrain with numerous rocks but is practically devoid of sharp, open fractures. Some lobate sub-units, similar in form

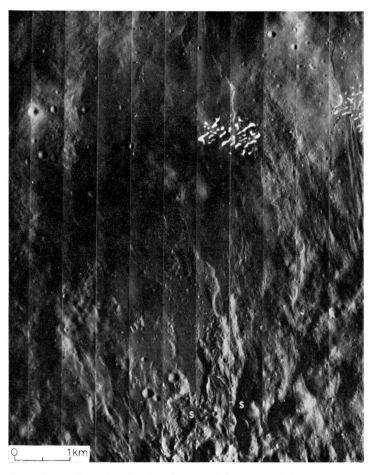

Fig.27. Thin, smooth flow units on the northern rim of Aristarchus. Possible sources of the flows are indicated by *S*. (Orbiter 5 photo, NASA.)

Fig.28. Narrow flows and associated drainage channels *(f)* on the northern rim of Aristarchus. Several of the flows have originated from the lake *(L)*. Another flow has issued from the flat-floored crater *(C)*. (Orbiter 5 photo, NASA.)

to units Af_{1-9} but usually thicker and coarser in texture, are found at the northern margins of unit Aa. The sub-units are included, here, in unit Aa, although it is not entirely clear whether or not they are younger than unit Aa and should be regarded as separate units.

Unit Ab is smooth, hummocky and devoid of open fractures. It is overlain by the other deposits and is therefore the oldest unit mapped. It is similar to the Tycho unit Tc. This deposit contains numerous concentric ridges which may be base surge dunes resulting from an explosion centred in Aristarchus.

Lakes similar to those surrounding Tycho are also found on the rim of Aristarchus. However, they are less numerous and many seem to define the source of the more recent flows. Several have well defined fronts which give a bladder-like appearance to the lake *(B* in Fig.29); others have numerous circular to irregular depressions similar to collapse depressions on terrestrial lava flows (Fig.32). Lakes which have given rise to flows appear to have overflowed their confining basins and then subsided to lower levels, in some cases leaving small tongues of flows (Fig.29) at higher elevations. Lakes also occur in low areas between fault scarps on the interior wall of Aristarchus, and these lakes have most of the above characteristics. Several flows (Fig.33) have originated at high levels and flowed downhill to partially or entirely fill a low area in the wall and form a typical lake. Closely spaced, open fractures parallel to the crater wall are common and indicate that a slight slumping of the wall took place after the formation and consolidation of these lakes.

Numerous flows, in addition to those previously mentioned, occur elsewhere on the rim of Aristarchus. Several examples are shown in Fig.34 and Fig.35. A series of fourteen short flows (Fig.34) have issued from small, circular to elliptical, craters on the southwestern rim. The flows range from 140 m to 1.4 km long and 50–390 m wide, and have a surface texture similar to that of the flows on the northern rim. A long, narrow flow (Fig.35), 4.4 km long and 28–570 m wide, originated from a lake L on the southwestern rim and flowed in a southeasterly direction across the grain of

Fig.29. Narrow flows *(f)* on the northern rim of Aristarchus. The lake *(A)* has apparently overflowed to form the flows and then subsided to a lower level. Notice the lake *(B)*, with a bladder-like form, occupying the shallow depression. (Orbiter 5 photo, NASA.)

Fig.30. Flows *(f)* which have been erupted from a trough or row of vents *(V)*. The arrows show the direction of flow. (Orbiter 5 photo, NASA.)

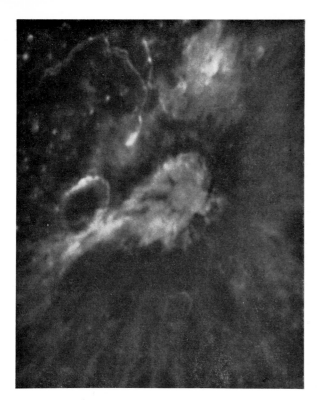

Fig.31. Full-Moon photograph of Aristarchus. The smooth flows Af_1–Af_9 occur within the dark area on the northern rim of the crater which has an albedo similar to that of the maria. The dark band on the eastern rim coincides with a row of lakes. (Catalina Observatory photo, Consolidated Lunar Atlas.)

Fig.32. A lake, on the eastern rim of Aristarchus, with a bladder-like form and containing numerous collapse depressions (C). (Orbiter 5 photo, NASA.)

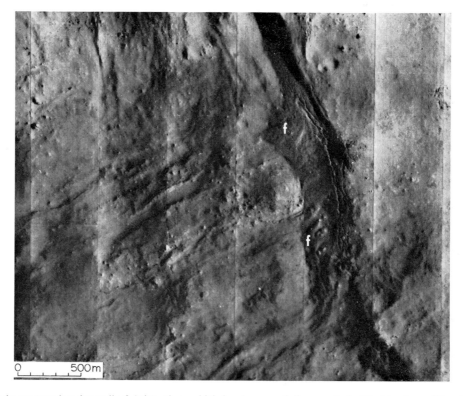

Fig.33. Lake, on the western interior wall of Aristarchus, which has been partially or entirely filled by flows (f) originating at higher levels on the wall. (Orbiter 5 photo, NASA.)

Fig.34. Flows *(F)* which have been erupted from small craters on the southern rim of Aristarchus. (Orbiter 5 photo, NASA.)

the underlying deposit. It formed a channel *C* with raised rims along the first one-third of its length. The portion of the flow where it emerges from the channel displays arcuate flow ridging *(r)*, and, at the middle and near the end, there are several circular, irregular, and strings of subdued craters *(d)* with convex inner walls (dimple craters) which resemble collapse depressions. The volume of the flow is considerably greater than the depression from which it issued, indicating that the flow was sustained by a continuous eruption. This is also true for many of the other flows.

As in the case of Tycho, the inner walls of Aristarchus also show evidence of flows. The flows down the walls of Aristarchus are expressed primarily as channels with raised rims. Many of the rims, or levees, originate at irregular craters and, in some cases, flows emanate from the distal ends of a channel (Fig.36). Fig.37 shows a flow which has issued upon the floor of Aristarchus and travelled about 840 m across the

floor. The surface texture of this flow is similar to that of the floor, but smoother, indicating that the material composing the flow and the floor is similar. One channel deposit has fractured in a way that characterises the cooling of a molten material.

It has been suggested (CRITTENDEN, 1968) that the flows are "mudflows" resulting from emanations of fluid (presumably liquid water or gas), and that the channels on the interior walls of Aristarchus and Tycho are debris channels caused by local slides. However, it would be difficult to account for the enormous quantities of water and its incorporation into the deposits necessary to explain these units as mudflows.

The present interpretation of the flows and lakes as highly fluid basaltic lavas is consistent with the morphology and the time sequence of deposition of the flows and lakes. Furthermore, the presence of collapse depressions on several flows and lakes, the association of flows with craters and lakes, and the

Fig.35. Long, narrow flow which originated from the lake *(L)*. The flow has formed a channel *(C)* along the first one-third of its length, and contains arcuate flow ridging *(r)* where it has issued from the channel, and collapse depressions *(d)* near its end. (Orbiter 5 photo, NASA.)

Fig.36. Thin, smooth flow *(F)* which has issued from a channel on the northern interior wall of Aristarchus. The flow apparently started from the flat-floored crater *C*. (Orbiter 5 photo, NASA.)

non-random distribution of the flows, are strong indications of volcanic activity. On the other hand, evidence for slumping of the interior walls of Aristarchus is considerable, and it is therefore probable that some of the channels are caused by debris slides.

The floor of Aristarchus exhibits a ropy, or ridged, texture with numerous open fractures, blocks and small hills. It is similar to the floor of Tycho, but portions of the floor are topographically more subdued than others. The floor unit Ap_1 (Fig.38) is characterised by an abundance of open fractures, a high crater density relative to the other floor units, and a fresher appearance. Unit Ap_3 is similar to unit Ap_2 but is relatively smooth and largely devoid of hills. The subdued nature of units Ap_{2-3} and the paucity of craters relative to unit Ap_1 indicate that portions of the floor have been partially covered by material emplaced after the deposition of unit Ap_1. This material may be volcanic ash erupted from the numerous hills, many with craters in their summits, which are found on unit Ap_2. Infrared and ultraviolet photography by E. A. Whitaker

(unpublished data, 1967), and photometric measurements by GEHRELS et al. (1964), have shown that the central portion of the floor of Aristarchus is bluish. This is surrounded by a peripheral zone of less blue material. The bluish material is apparently represented by the blanket forming units Ap_2 and Ap_3, and less blue material seems to coincide with unit Ap_1.

Unit Ap_1 contains a well defined, elevated bench along most of the margins of the floor. This bench (Fig.39) contains numerous open fractures which are concentrated in the area of maximum curvature and indicate that the floor has subsided leaving an elevated bench broken by tension fractures in the zone of maximum downbending. The floor of Tycho has a similar, but less pronounced, bench. This feature is strikingly like the fractured benches (Fig.40) found on terrestrial lava lakes caused by the drainback of lava into the vent. If the Aristarchus bench formed in this manner, the amount of material which drained away can be estimated from the height of the bench and the diameter of the floor. Using a minimum floor diameter of

Fig.37. Flow *(f)* which travelled down the western interior wall of Aristarchus to form a channel *(c)* and spread across the floor. (Orbiter 5 photo, NASA.)

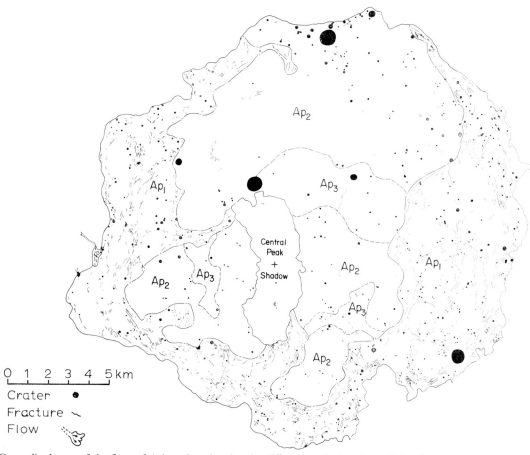

Fig.38. Generalised map of the floor of Aristarchus showing the different geologic units and the distribution of craters and fractures. (Compiled from Orbiter 5 photo.)

Fig.39. A portion of the southern floor of Aristarchus at its contact with the wall *(W)* showing a relatively thin peripheral bench *(B)* with open fractures in the zone of maximum downbending *(D)*. (Orbiter 5 photo, NASA.)

12 km and an estimated minimum bench height of 25–50 m (from shadow measurements on the eastern floor), the minimum volume of material which has drained away is of the order of 3–6 km³. However, estimates based on the fracture width and on the assumption that there has been no slumping of the sides of the fractures, yield a bench height and, therefore, volume of material drained away, an order of magnitude greater. In any event, if the bench were formed by the subsidence of the floor due to the drainback of an appreciable volume of lava, both Aristarchus and Tycho—and other craters exhibiting similar benches—would probably be underlain by chambers or open fissures capable of expelling and receiving large volumes of magma.

Crater counts in and near to Aristarchus

The measured areas—corrected for shadowed and overexposed areas—of the different units and counts of all circular, eumorphic craters down to 28 m in diameter in these units are given in Table IV. The data are represented logarithmically in Fig.41.

Many of the units mapped around Aristarchus offered such a small surface that individual counts could not be regarded as significant. In particular, the data for units Af_2–Af_8 were combined in order to obtain a "significant" counting area (18 km²).

Specific comparisons of the cratering of the various units may be made for craters of minimum size 56 m: this size group may represent the most reliable data

Fig.40. Bench at the margin of the floor of Kilauea Iki crater (Hawaii), showing peripheral fractures. The bench resulted from the drainback of the lava lake into the vent. (Photo by D. P. Cruikshank.)

TABLE IV

TOTAL CRATER COUNTS
IN DIFFERENT GEOLOGIC UNITS OF ARISTARCHUS

Geologic unit	Corrected area of count (km^2)	Total number of craters $\geqslant 28$ m in diameter	Craters per unit area (floor unit $Ap_1 = 1$)
Ap_3	19	26	0.3
$Ap_2 + Ap_3$	85	193	0.5
Ap_1	83	409	1.0
Af_{2-8}	18	88	1.0
Af_1	17	104	1.2
Aa	65	573	1.8
Ab	58	247*	—

* Total number of craters $\geqslant 56$ m in diameter.

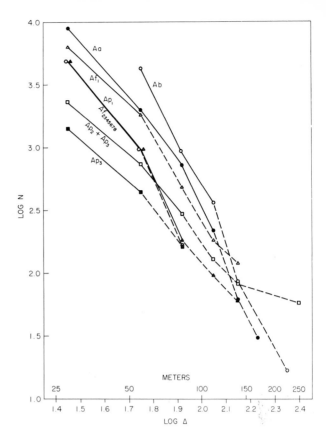

Fig.41. Logarithmic plot of the cumulative number *(N)* of craters per 1,000 km² of surface against minimum crater diameter (Δ) for the geologic units in Fig.26 and 38.

since, in Fig.41, there may be some losses for the smaller craters in all units. At the 56-m size there are 2.1 times as many craters per unit area on unit Ab as on unit Aa. There are only 1.1 times as many on Aa as on Af_1; but Af_1 overlies Aa. There are 2.1 times as many craters per unit area on Aa as in unit area of Af_{2-8}.

Considering the floor units (Fig.38) of Aristarchus, Ap_1 is progressively more heavily cratered than $Ap_2 + Ap_3$ for crater sizes less than 65 m but Ap_1 has a number density of craters larger than 65 m in diameter that is lower than that of $Ap_2 + Ap_3$. One possibility is that many of the smaller craters on units $Ap_2 + Ap_3$ have been obliterated, by mass wasting of hills present only in these units, by flows, or by pyroclastic deposits, and that the population curves are not significantly different for diameters in excess of about 65 m. Another possibility is that the craters are members of inherently different populations; in which case many of them are probably of internal origin. Since the high-resolution photographs of the floor units $Ap_2 + Ap_3$ show a deficiency of open fractures, and give the impression that these units have been covered by some deposit, we adopt the first explanation.

In order to test the hypothesis that the blanketing materials were derived by mass wasting of hill slopes, we isolated those regions (Ap_3) of unit $Ap_2 + Ap_3$ which were largely devoid of hills, and performed counts of craters in the unit Ap_3 alone. The counts (Fig.41) indicate that the population of craters on Ap_3 is the same as on Ap_2. In addition, there are more craters per unit area of Ap_2 than there are per unit surface of Ap_3—just the opposite of what one would expect if mass wasting of hill slopes were the operative mechanism obliterating craters. Therefore, the blan-

keting is probably the result of flows and/or pyroclastic deposits.

It should also be noted from Fig.41 that the floor unit Ap_1 is cratered to the same number density as the rim flows Af_{2-8}. This indicates that, on average, the oldest floor unit, Ap_1, was emplaced at about the same time as most of the mapped rim flows. This is not the case for Tycho, in which the mapped rim flows were apparently in place long before the floor formed.

In most cases, there are more craters per unit area of stratigraphically older units. This conclusion holds for all the sizes (28–112 m) covered by the more significant points of Fig.41. This indicates that the majority of the eumorphic craters counted are of either primary or secondary impact origin.

Facts relevant to the interpretation of Aristarchus

In the following summary, reference may be made to Fig.26–41.

(1) Counts of craters larger than 56 m in diameter

show that stratigraphically younger units of Aristarchus carry fewer craters and that, therefore, the majority of the craters counted are impact craters.

(2) The fact that there are essentially the same number of craters per unit area on the combined flows Af_{2-8} as on the fractured part (Ap_1) of the floor of Aristarchus indicates that the flows are, on average, of the same age as the fractured floor unit.

(3) The curves (Fig.41) for the fractured floor unit Ap_1 and for the unfractured floor units $(Ap_2 + Ap_3)$ intersect in their more significant portion. This shows that Ap_1 has more craters smaller than 65 m in diameter than $(Ap_2 + Ap_3)$; and, conversely, $(Ap_2 + Ap_3)$ has more craters larger than 65 m in diameter.

(4) Certain portions of the floor have been designated Ap_3 since they are relatively free of hills and are much smoother than the remainder of the floor (Ap_2). Unit Ap_3 carries 1.9 times fewer craters than Ap_2, yet it maintains the same slope as Ap_2 in Fig.41, and this slope is sensibly less than that of any of the other units.

(5) Many flows occur in the Aristarchus region. They are up to 9 km long and 400–4,000 m or more wide. The widest flows may in reality be composed of coalescing flows. Some flows are exceptionally long relative to their width.

(6) There are flows on the inner wall of Aristarchus. One reached and crossed part of the floor unit and so was later than that part of the floor.

(7) Near the southern rim of Aristarchus there are at least fourteen craters each with a flow emanating from it: each flow is directed away from Aristarchus, apparently downslope. In another instance, a flow has started in a "lake" and flowed downhill tangentially to the rim of Aristarchus. This has been confirmed using stereo photographs.

(8) Some "lakes" are found in the vicinity of Aristarchus, and several have well developed fronts with numerous collapse depressions.

(9) There is an inward-sloping bench around the margins of the floor of Aristarchus. The bench contains open fractures, characteristic of a zone of maximum downbending.

(10) Different units have marked differences in texture. The floor of Aristarchus is generally similar to that of Tycho except in so far as that no open fractures are found in the floor of Aristarchus other than in the peripheral parts. Flows Af_{1-8} have smoother surfaces than the other flows and their albedo is approximately the same as that of the maria.

Comparison of Tycho and Aristarchus

Flows and lakes

Most of the flows associated with Tycho have morphologies and structures that differ from those associated with Aristarchus. The Aristarchus flows are thinner, smoother, have more highly developed lobate margins and are not as bright as the Tycho flows. The flow designated C (Fig.13, 14) near Surveyor 7 is similar to those in the vicinity of Aristarchus in that it is relatively smooth and thin. However, even this flow is more ridged and has a higher albedo than the Aristarchus flows. Furthermore, many of the Aristarchus flows have issued from lakes whereas only several minor flows in the vicinity of Tycho are due to the overflow of material from the lakes. All the above facts strongly suggest that, relative to the Tycho flows, the Aristarchus flows were considerably more fluid. They probably consist of basaltic lava typical of the maria whereas the Tycho flows, of intermediate albedo, represent a more differentiated material that may be higher in silica. This is not unexpected, since Aristarchus is situated in mare material and Tycho is in the highlands.

The morphology of the lakes associated with both Tycho and Aristarchus is similar. However, the lakes in the vicinity of Tycho are generally larger, more numerous, and lie in depressions rather than give rise to flows. Several of the Aristarchus lakes have been the sources of flows, and several lakes exhibit an abundance of collapse depressions which are not seen on the Tycho lakes. In general, however, the lakes in both areas seem to have originated by the same process.

Floor units

The gross features of the floors of Tycho and Aristarchus strongly suggest that, in each case, the material solidified from a melt. However, the floor of Tycho has not been subjected to the blanketing deposits present in the central portions of the floor of Aristarchus. It would seem that, after Aristarchus was partially filled with lava and the surface solidified, the still molten lava beneath the solidified crust drained back into a chamber or open fissures to leave a peripheral bench. This was followed by the eruption of ash, possibly from the numerous hills on the floor, which blanketed or subdued the pre-existing floor structure. The fact that the central portion of the floor of Aristarchus is bluish and is surrounded by a less blue peripheral zone supports this view.

Interpretation of crater counts

Base surge hypothesis

A base surge is a ring-shaped gas-charged density flow formed by expanding gases resulting from an explosion. The surge cloud is propelled over the crater lip by the expanding gases and carries large amounts of ejecta in radial directions away from the crater and nearly parallel to the ground surface. The cloud may travel at initial velocities of over 50 m/sec and may carry fragmental material over many kilometres. Base surges may result from artificial, meteoric or volcanic explosions.

MOORE (1967) described the mechanics of base surge formation from an artificial explosion as follows: "In an underground shot expanding gases at the explosion centre first vent vertically, pushing up and out a 'wall' of nearly coherent roof material. Only after the wall is broken and bent down and outward can expanding gases from the explosion crater rush over it, erode it, and carry material from it outward to feed the base surge. Trajectories of previously ejected material are necessarily at a high angle because of the directing effect of the wall. The early falling, larger blocks may strike in front of the base surge after it has begun to form. Such blocks are quickly overrun by the surge cloud and most other throwout material comes down later and falls into the moving cloud". Although this description is of an artificial explosion, explosive volcanic eruptions react in like manner. In both cases the base surge deposits concentric dunes having the same centre as the crater rim.

It has been suggested privately to the authors that the deposits surrounding Tycho and Aristarchus are base surge deposits and that the vast majority of craters on the various units are secondary impact craters formed on the units soon after their deposition. According to this hypothesis, the differences in crater frequency between the different units resulted from the decreasing amounts and higher trajectories of ejecta as the younger deposits were laid down. Although we believe that certain units surrounding Tycho and Aristarchus are base surge deposits, there are serious objections to several of the units being the result of a base surge and strong evidence indicating that they are volcanic flows.

It was shown above that the morphology of the flows Tf_1 and Tf_2 associated with Tycho is similar to that of known viscous lava flows on Earth. In discriminating between base surge deposits and lava flows there are three additional points to note: (1) the flows

appear to be gravity-controlled so that, in particular, a few of them flow in directions other than away from the rim of Tycho and Aristarchus; (2) others definitely originate in, and emerge from, breached craters which themselves are located on, or outside, the walls of Tycho and Aristarchus; and (3) the excess of crater pits counted by one of us (R.G.S.) on several flows near to Surveyor 7 may best be explained as volcanic explosion craters. All three points demonstrate that the flows in question cannot be base surge deposits.

A feature of the base surge hypothesis is that it predicts that the bulk of the recognisable secondary craters in a particular surge deposit formed immediately after the formation of that deposit: in this way, stratigraphically recent units are expected to be less heavily cratered than the earlier units. Furthermore, the floor of the crater from which the base surge evolved is expected to be the least cratered of all the units. This, indeed, is the situation revealed by observations of Tycho. By contrast, in the case of Aristarchus it is found that the floor unit Ap_1 is cratered to the same extent as units Af_{2-8} exterior to Aristarchus (see Fig.41). Thus, again, the conclusion is that the base surge hypothesis is inadequate, by itself, to explain the facts.

Finally, the "duning" or ridging that is approximately concentric with Tycho and Aristarchus, and appears outside their confining walls, has been cited as one of the effects of a base surge, and it does seem probable that such duning is present. As the dune-like structures are traced back to the crater rims at several localities near the rims of both Tycho and Aristarchus they become progressively more scarp-like until, very near the rims, they show definite displacements of the surface. This indicates that at least much of the concentric ridging near the crater rims is fracturing resulting from subsidence or slumping of the crater rim and surroundings. The ridging is particularly well developed around Aristarchus, but parts of it trend tangentially to the wall segments of Aristarchus and Tycho to form complex lattice patterns. Although much of the concentric ridging—particularly the more distant ridging—is probably base surge duning, we believe that the complex lattice pattern is the result of faulting in the outer walls of the large craters; possibly lavas issued from some of these fractures.

Whilst certain of the mapped units are almost certainly congealed lava flows, and the majority of the units mapped are probably volcanic flows, it does appear possible that other units—for example, Tc and Ab—are base surge deposits.

Dating Tycho and Aristarchus

On the base surge hypothesis it is clear that the crater counts may be compositions of at least two populations of craters deriving from meteoroids and from secondary ejecta. Absolute dating based on counts of eumorphic craters will therefore depend on the validity of the assumption that the craters are in volcanic flows rather than deposits from base surges. A third population, consisting of endogenic craters, does not appreciably influence the counts of the larger craters.

With these assumptions, dating is performed by adopting the particle influx law which best represents all observations relating to the minor objects compiled by VEDDER (1966) who proposed fitting them by means of eq.1:

$$\log I = -14 - \log m \qquad (1)$$

where I is the total number of particles of mass greater than m gram crossing one square meter of cis-terrestrial space every second. It is assumed that, for the small, eumorphic craters counted in the present exercise, the same law of accumulation holds, at the present epoch, for the Moon.

ÖPIK (1962) has given equations that relate the diameter of an impact crater to the diameter of the parent meteoroid for different particle densities and velocities of impact. With a representative diameter of $\Delta/15$ for a metoroid of density 3 g/cm^3 which forms a crater of diameter Δ, it follows that the mass required to form a crater 50 m or more in diameter is $5.8 \cdot 10^4$ kg. Under these conditions, eq.1 predicts one primary lunar impact crater 50 m or more in diameter per 1,000 km^2 of surface every $1.8 \cdot 10^5$ years.

Using the cumulative number-data of Table V and VI, it follows that the various geologic units may be expected to have the ages listed in the last column of each table

The spread of ages is large for both Tycho and Aristarchus. If the estimates are taken at face value, the time span between the formation of the Tycho lavas $Tf_1 + Tf_2 + Tf_3$ and $Ta + Tb$ is of the order of 10^8 years; and, between the Aristarchus flows Af_{2-8} and Af_1, it is $1.5 \cdot 10^8$ years. The floor of Tycho and the lava lakes solidified $1.3 \cdot 10^8$ years later than the lava flows $Tf_1 + Tf_2$. The floor of Aristarchus itself developed over a period of 10^8 years and, unlike Tycho, parts of the floor of Aristarchus appear to be 10^7 years older than the flows Af_{2-8}. The flows Af_{2-8} are differently cratered even among themselves and probably developed over widely different times. Hence the in-

TABLE V

CUMULATIVE CRATER COUNTS
IN THE DIFFERENT GEOLOGIC UNITS OF TYCHO

Unit	N^*	Age of unit (in 10^8 years)
Tl	759	1.4
Tp	869	1.6
$Tf_1 + Tf_2 + Tf_3$	1,575	2.8
$Ta + Tb$	2,005	3.6
Tc	3,871	7.0

* N is the number of craters $\geqslant$ 50 m in diameter per 1,000 km^2.

TABLE VI

CUMULATIVE CRATER COUNTS
IN THE DIFFERENT GEOLOGIC UNITS OF ARISTARCHUS

Unit	N^*	Age of unit (in 10^8 years)
Ap_3	501	1.0
$Ap_2 + Ap_3$	841	1.6
Af_{2-8}	1,259	2.3
Ap_1	1,349	2.3
Af_1	2,138	4.0
Aa	2,630	4.6
Ab	7,763**	11.0

* N is the number of craters $\geqslant$ 50 m in diameter per 1,000 km^2.
** Value found by extrapolating best fitting line.

dication that some are more recent than parts of the floor of Aristarchus is strengthened. The scatter in the original data that were used to construct eq.1 is such that the deduced ages might be in error by a factor of 10 either way. GAULT and GREELEY (1968) have proposed an age of as little as 10^6–10^7 years for Tycho, but have used a flux that is substantially higher than that given by eq.1. It is considered that the bulk of the observations that led to the formulation of eq.1 cannot be waived so lightly. Even if the ages proposed here were in error by an order of magnitude, there would still be wide intervals of time between the formation of the various lava fields. It would, of course, be surprising were this not so. Therefore, it seems that multiphase eruptions characterised at least part of the formation of the structures Tycho and Aristarchus.

The development of Tycho and Aristarchus may be compared by reference to Table VII.

The floor and flow units are of the same order of age, the most recent features being the floor of Tycho and the central parts of the floor of Aristarchus. The flows appear to have started first in the case of Aristarchus and the last ones appear to have been later than those of Tycho. The biggest difference in the crater

TABLE VII

COMPARISON OF AGES OF THE GEOLOGIC UNITS
ASSOCIATED WITH TYCHO AND ARISTARCHUS

	Proposed age (in 10^8 years)	
	Tycho	Aristarchus
Floor	1.6	1.0–2.3
Flows	2.8–3.6	2.3–4.0
Principal external unit	7.0	11.0

densities is encountered in the principal external units Tc and Ab. If the craters counted were mostly primary impact craters, the terrain around Aristarchus would be 10^9 years old, and 1.6 times as old as the terrain around Tycho.

Conclusions

Both Tycho and Aristarchus probably originated in the same manner, although there were differences in their later development. Extensive lava flows issued from the inner and outer rims of both craters. Flows from small craters in the vicinity of the Tycho and Aristarchus structures show that these small craters are volcanic. This establishes that one class of lunar volcano, at least, resembles an impact crater.

Confirmed observations of red glows and other transient phenomena on the rim of Aristarchus and vicinity (ANONYMOUS, 1964; GREENACRE, 1965; HARTMANN and HARRIS, 1968), as well as unconfirmed reports of similar phenomena in Tycho (MIDDLEHURST et al., 1968), suggest that limited volcanic activity has continued to the present day.

Evidence of the drainback of lava and consequent subsidence of the floors of Tycho and Aristarchus suggest that both craters are underlain by chambers or extensive networks of fissures capable of receiving large volumes of magma. The evidence is for the generation of magma at more than superficial depths. Magmatic differentiation seems to be required to account for the more viscous flows and, in particular, the floor lavas of both structures. This is supported by the composition of samples collected by the Apollo 11 and 12 astronauts. In the case of Aristarchus, very late eruptions appear to have been responsible for obscuring certain parts of the floor detail; this material may be lavas and/or pyroclastic deposits.

Three possible causes of the volcanism associated with Tycho and Aristarchus have been suggested: (1) sufficient thermal energy was produced by an impact

to generate a large pocket of magma which gave rise to the volcanism; (2) the volcanism was triggered by an impact that penetrated, or produced fractures which tapped, a potential source of magma; and (3) Tycho and Aristarchus are volcanic structures and the observed volcanism represents the natural sequence of development of the craters.

It seems doubtful that the amount, distribution and temporal relationships of the volcanic units can be accounted for on the basis of impact generated volcanism. First, if the floor lavas were generated by the thermal energy of an impact, they should be the same age as, or older than, the rim units. However, the crater counts show that the number density of craters on the floors is considerably less than that of most of the rim units, and, therefore, that the floors were emplaced at a much later time than the rim units. Second, volcanism generated by an impact would be expected to cease within a relatively short time, and would not be expected to continue intermittently over periods of millions of years. In the case of Aristarchus, volcanic activity appears to have continued to the present time (HARTMANN and HARRIS, 1968). Third, volcanism associated with Tycho occurs at distances of at least 100 km from the rim, and it is unlikely that an impact of the size necessary to produce Tycho could generate volcanic activity at such a great distance from the parent crater. Fourth, a floor melt produced by impact would seal off subsurface fractures almost immediately, and it is unlikely that large volumes of magma could drain away as indicated at both Tycho and Aristarchus.

The floor of a recent impact crater would tend to adjust upwards—not downwards, as observed—under the influence of isostasy: the peripheral fractures in the downwarped benches of the floors of the structures, even if originally present on an impact hypothesis, would probably close as a result of the isostatic uplift of the lowest parts of the floors after some 10^8 years had elapsed (FIELDER, 1965)—that is, in a time that was comparable to the age of either structure. Again, craters which have been completely filled, such as Wargentin (essentially the same size as Tycho), could not possibly have had so much lava generated by an impact. The floor material of such craters must have originated from a relatively deep-seated and prolific source of lava.

It is possible to explain the observed temporal and spacial relationships of the volcanism associated with Tycho and Aristarchus by an impact triggering mechanism if it is assumed that a potential zone of magma occurred at a relatively shallow depth beneath the lunar surface. A "potential" source is one that may become an active source upon a change of physical parameters, such as pressure and temperature. In this case, an impact could produce fractures which tap a potential subsurface source of magma and give rise to volcanic activity within a crater and probably at considerable distances from the rim crest. This volcanism might occur intermittently and persist for a considerable length of time. It is conceivable that a large impact generating Tycho or Aristarchus could produce fractures deep enough to give rise to volcanism.

If Tycho and Aristarchus were purely volcanic structures, the observed volcanism would be a natural consequence of the development of the craters. If the relatively old units Tc and Ab were base surge deposits their features would have been sculptured by explosions centred in the respective craters Tycho and Aristarchus. The extensive ray systems, numerous large blocks on the rim, and swarms of probable secondary impact craters at distances of several diameters from the parent craters form additional evidence for large explosions. Although base surge deposits are characteristic of explosive volcanic eruptions, it is problematical whether enough internal energy could be concentrated in a relatively small area to produce such large explosions. Irrespective of the initial mode of formation of Tycho and Aristarchus, impact or volcanism, there is compelling evidence for subsequent, widespread volcanism associated with each crater and covering a protracted time scale.

References

ANONYMOUS, 1964. *Aeronaut. Chart. Inform. Center, Tech. Papers*, 12.

CRITTENDEN, M. D., 1968. *Langley Res. Center, Working Papers*, 506: 158.

FIELDER, G., 1965. *Lunar Geology*. Lutterworth, London.

GAULT, D. E. and GREELEY, R., 1968. *Trans. Am. Geophys. Union*, 49: 273.

GEHRELS, T., COFFEEN, T. and OWINGS, D., 1964. *Astron. J.*, 69: 826.

GREENACRE, J. C., 1965. *Ann. New York Acad. Sci.*, 123: 811.

HARTMANN, W. K. and HARRIS, D. H., 1968. *Commun. Lunar Planetary Lab. Univ. Arizona*, 7: 161.

MCCONNEL, R. K., MCCLAINE, D., LEE, W. D., ARONSON, J. R. and ALLEN, R. V., 1965. *Rev. Geophys.*, 5: 121.

MIDDLEHURST, B. M., BURLEY, J. M., MOORE, P. and WELTHER, B. L., 1968. *NASA, Tech. Rept.*, TR-R-277.

MOORE, J. A., 1967. *Bull. Volcanol.*, 30: 337.

ÖPIK, E. J. 1962. In: *Progress in the Astronautical Sciences*. North-Holland, Amsterdam, 1: 219.

PATTERSON, J. H., TURKEVICH, A. L., FRANZGROTE, E. J., ECONOMON, T. E. and SOWINSKI, K. P., 1970. *Science*, 168: 825.

PECK, D. L. and MINAKAMI, T., 1968. *Geol. Soc. Am., Bull.*, 79: 1151.

PETTENGILL, G. H. and THOMPSON, T. W., 1968. *Icarus*, 8: 457.

SAARI, J. M. and SHORTHILL, R. W., 1965. *Nature*, 205: 964.

VEDDER, J. F., 1966. *Space Sci. Rev.*, 6: 365.

WOOD, J. A., DICKEY, J. S., MARVIN, U. B. and POWELL, B. N., 1970. *Science*, 167: 202.

6

Geology of the farside crater Tsiolkovsky

J. E. GUEST

Introduction

Tsiolkovsky (22 °S 128 °E) forms a striking feature (Fig.1–3) on the densely cratered highlands of the lunar farside, being readily recognised by its dark, flooded floor. Although it has a diameter of about 200 km it has also many of the features typical of smaller craters such as Copernicus, Aristarchus and Tycho and is probably of the same origin. Again, Tsiolkovsky compares with the circular, flooded maria. A number of authors have suggested that Tsiolkovsky forms a link between large craters and the circular maria.

The age of Tsiolkovsky, relative to the lunar stratigraphic column, has not been derived directly by geologic mapping. Nevertheless, from a study of the denudation chronology of lunar craters, OFFIELD and POHN (1969) concluded that Tsiolkovsky was younger than Mare Orientale but older than most of the dark surface units of the mare basins. Probably Tsiolkovsky belongs to the Imbrian System in the lunar stratigraphic time scale, and is equivalent to craters of the Archemedian Series. Offield and Pohn gave the age of the mare material in Tsiolkovsky as being about the same as that of the surface units in eastern Oceanus Procellarum.

GUEST and MURRAY (1969) defined and mapped the principal geologic units of Tsiolkovsky. Evidence involving the morphology of the crater, the nature of an associated ejecta blanket, and the presence of a well defined annulus of apparent secondary impact craters outside the thickest development of ejecta, led them to conclude that Tsiolkovsky resulted from a single, high-energy explosion[1]. Later erosion and volcanic activity modified the crater to its present form.

---

[1] The term "explosion" is used in an informal sense throughout this chapter.

The crater

The outline of the lip (Fig.4) of Tsiolkovsky is polygonal, indicating that its shape was controlled by an underlying structural pattern. However, GUEST and MURRAY (1969) showed that, like many other large lunar craters, it had a high degree of circularity typical of craters formed by a single explosion. Craters formed by multiple explosions (for example, most volcanic craters) or by caldera collapse are much less circular (MURRAY and GUEST, 1970). This is considered to be an important criterion for identifying the origin of this crater and others like it.

The rim, also considered to provide strong evidence regarding the origin of Tsiolkovsky, is defined here as the material forming the raised feature outside the apparent crater, from the lip to the main break in slope at the foot of the outer walls. BALDWIN (1963) showed that there was a relation between the rim widths and diameters of lunar craters. GUEST and MURRAY (1969) extended Baldwin's data to include: (a) a wider spread in the diameters of lunar craters; and (b) volcanic craters of different types. It is clear from Fig.5 that Baldwin's relation holds for a large number of lunar craters of all sizes, including Tsiolkovsky. A similar relation also holds for single explosion craters such as meteoritic craters and nuclear explosion craters. Most volcanic craters—particularly calderas— have much wider rims.

This difference is thought to result from the respective mechanisms of construction of the rims in the two cases. In a volcanic crater the rim is built up by accumulation of either pyroclastic material or lavas, and the width of rim depends on the distance travelled by the erupted products.

The rims of high-energy explosion craters, including high-velocity impact craters, are formed mainly by the uplift and overturning of country rock by the

Fig.1. Tsiolkovsky and the surrounding highlands of the lunar farside. Note the large, denuded crater cut by Tsiolkovsky on its western side. (Orbiter 3 photo, NASA.)

Fig.2. High-resolution Orbiter 3 photograph of the northwest part of Tsiolkovsky. A geologic sketch map of this area is given in Fig.7
(NASA.)

explosion, only a small part of the total thickness of the rim consisting of ballistically ejected particles. It may be that the rim widths of Tsiolkovsky and other lunar craters of this type represent a zone of intense structural deformation and uplift. Certainly the strong morphologic similarity between Tsiolkovsky, on the one hand, and terrestrial meteoritic and other high-energy explosion craters, on the other, adds evidence to the concept that Tsiolkovsky is an impact crater.

Terracing—a feature characteristic of almost all lunar craters larger than about 20 km in diameter—is well developed on the inner walls of Tsiolkovsky. The terraces are formed by a number of slumped or down-faulted blocks controlled by fractures concentric with the lip of the crater. There is some tendency for this faulting to follow regional trends. The features of the terraces are somewhat rounded and appear to have suffered some erosion; and the presence of a fine ridge texture, suggestive of creep, implies that there has been a general maturing of the topography by a process of slow, downslope mass movement of the material. Similar features are observed on the rim outside the crater lip.

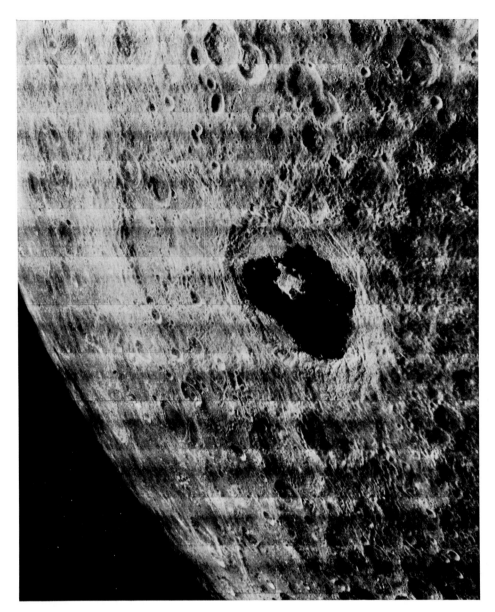

Fig.3. Oblique Orbiter 1 photograph of Tsiolkovsky from the northeast. (NASA.)

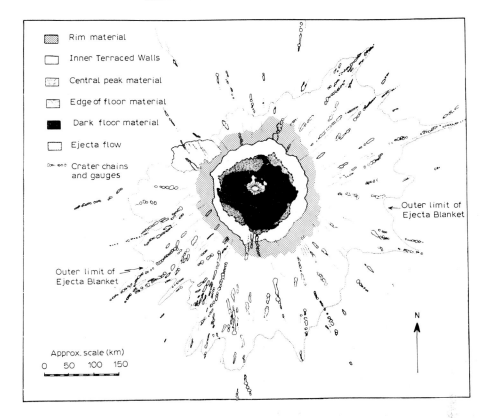

Fig.4. Geologic map of Tsiolkovsky constructed from Fig.1. (After GUEST and MURRAY, 1969.)

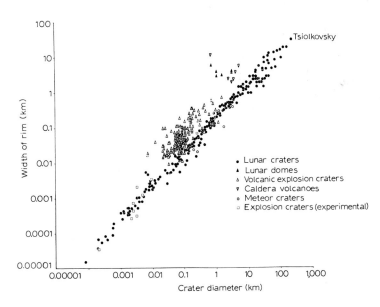

Fig.5. Relation of crater diameter to rim width for various lunar (from U.S. Aeronautical Chart and Information Center Charts and Orbiter and Surveyor photographs) and terrestrial craters. (After GUEST and MURRAY, 1969.)

The ejecta blanket

Surrounding the crater of Tsiolkovsky, and extending for some 200 km away from the crater lip, is a continuous mantle of fragmental material; this is considered to be ejecta, showing various stages of shock, thrown out from Tsiolkovsky. Its fragmental nature is attested to by the way it coats the underlying terrain, subduing the original morphology of older craters by burial. This burial of the underlying terrain to different degrees according to the thickness of the blanket has been observed for many of the Moon's large craters (for example, Tycho: SHOEMAKER et al., 1968).

The main part of the ejecta blanket (Fig.4) has a bilateral symmetry about a northwest–southeast axis through the centre of the crater. The ejecta blanket is taken to be direct evidence of an explosive origin of the crater. One possible cause of the bilateral symmetry is that an impacting body collided at an angle low enough to give the observed distribution of ejecta but not low enough to cause the crater to depart markedly from the circular form. The other possibility is that trajectories of ejecta were strongly controlled by structure in the excavated bedrock; the final distribution of ejecta would then reflect an underlying structural pattern.

The surface of the ejecta blanket is hummocky, and the largest hummocks are on the rim. In places near, and on, the rim the ejecta blanket is characterised by subradial ridges and valleys. Some of the hummocky terrain may be dune topography caused by a base surge (Chapter 5). The thickness of the blanket decreases away from the crater, some measure of the thinning being given by the degree to which the pre-existing terrain is buried.

Craters that are considered to be of secondary impact origin are well developed on and beyond the ejecta blanket of Tsiolkovsky. Radial and subradial crater chains become common in the outer parts of the ejecta blanket, replacing the characteristic ridges and grooves of the inner parts. The principal crater chains are marked in Fig.4. Some of the chains are over 170 km long. Craters in the chains tend to be elliptical, and adjacent craters overlap. Close examination of the overlapping relations of the craters shows that there is a tendency for the craters in a given chain to be younger as the distance from Tsiolkovsky increases.

In marial areas clusters of secondary craters associated with large, primary craters may be recognised readily since the background level of cratering is relatively low. In highland areas, such as that surround-

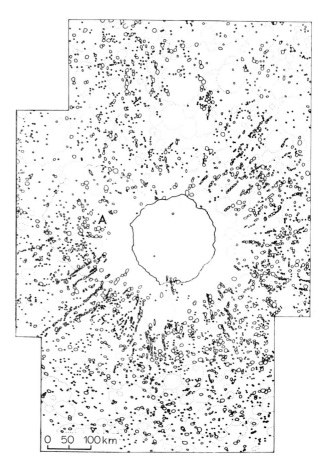

Fig.6. Map showing distribution of craters in the Tsiolkovsky area. Craters smaller than 10.6 km diameter are outlined with continuous lines.

ing Tsiolkovsky, secondary craters tend to be swamped by other craters. Fig.6 is a plot of all the craters in the Tsiolkovsky area. It can be demonstrated that craters above 10 km in diameter have an areal distribution which is independent of Tsiolkovsky and, indeed, many of them are older than the ejecta blanket. However, craters in the size range 3–10 km show a close spatial relation to Tsiolkovsky, forming an annulus of high crater density at about 120 km from the lip of the main crater. The distribution also conforms to the shape of the ejecta blanket, there being fewer craters on the northwest side where the blanket has its poorest development. It is concluded that the majority of the craters in this size range are secondary impact craters. This view conforms to the earlier evidence that Tsiolkovsky was produced by a large, single explosion.

The ejecta flow

This unit (*S* in Fig.2 and mapped in Fig.7) located on the northwest rim of Tsiolkovsky, forms a discrete mappable unit which differs in character and mechanism of emplacement from other units of the ejecta blanket. It is a wedge-shaped body, thinning away from the crater lip. In the lower part it has a flow-like form and the top surface is marked by longitudinal grooves some of which have widths of 400 m or more. The unit has a well marked front which is over 100 m high; it is irregular in form but is broadly lobate. Towards the rim of Tsiolkovsky the top surface of the unit becomes steeper, and the total thickness probably attains several hundred metres. GUEST and MURRAY (1969) pointed out that, high on the rim, the transverse ridges turned downslope near the edges of the unit rather than upslope which would have been expected had they been normal flow ridges. They concluded from this that the unit had formed by the collapse of the outer wall of the rim as a large landslide. The flat-surfaced flow-like continuation of the unit below this they considered to be produced by flow of debris away from the landslip collapse high on the rim. Inspection of Fig.7 shows that, adjacent to the supposed slide, the crater lip is offset outwards possibly as a result of wholesale outward displacement of the rim, causing the collapse of the outer rim.

It is of interest to note that the Sherman Glacier landslide (Fig.8) has many of the characteristics of this unit of Tsiolkovsky: notably, the well marked longitudinal grooves and the steep, irregular front. SHREVE (1966) concluded that the Sherman Glacier slide developed sufficient momentum, in the initial fall, for the sliding material to leave the ground at the first break in slope, trapping air below and allowing the

Fig.8. The Sherman Glacier landslide, south central Alaska. (U.S. Geological Survey photo.)

material of the landslide to travel forward for some considerable distance on a cushion of air.

The evidence from the Tsiolkovsky ejecta flow suggests that, like the Sherman Glacier slide, it was initiated by the bulk sliding of material on a slope. The morphologic similarity between the rock units developed at the foot of both these slides suggests that they were emplaced from slides generated by the same mechanism; that is, that the mode of transportation was the same.

Water, to act as a lubricant, was lacking in both cases. However, gas might not have been readily available to form a cushion on which the lunar slide could have been transported. There may, of course, have been a temporary atmosphere generated by the Tsiolkovsky explosion. Even if a direct analogy with the Sherman Glacier slide is not correct there seems little doubt from the evidence that the Tsiolkovsky ejecta flow originated as a landslide; it might perhaps be more appropriate to term this unit an *ejecta slide*.

Shreve considered the longitudinal grooves to have been formed by shear between substreams of debris advancing at different speeds. The longitudinal grooves on the Tsiolkovsky slide probably originated in the same manner. It is of interest that the grooves of the Tsiolkovsky unit not only strike in the same direction as the flow but also correspond to a well marked

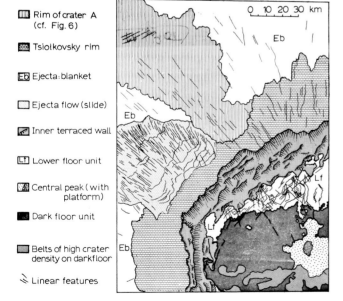

Rim of crater A (cf. Fig. 6)

Tsiolkovsky rim

Eb Ejecta blanket

Ejecta flow (slide)

Inner terraced wall

Lower floor unit

Central peak (with platform)

Dark floor unit

Belts of high crater density on dark floor

Linear features

Fig.7. Geologic sketch map explaining Fig.2.

regional structural trend of this area; possibly pre-existing fractures were utilised in differential shear movements within the slide during transportation.

Lower crater floor unit

Two distinct rock units outcrop on the crater floor. The younger of these is the dark, mare-like material which floods most of the floor but, at the edge of the crater floor, areas of the lower unit (Fig.7) remain uncovered. This lower unit forms a discontinuous bench at the foot of the inner, terraced walls and was, for this reason, termed the *edge of floor* material by GUEST and MURRAY (1969). The age relation between this unit and the dark floor unit is demonstrated by geometry of the contact between them, the dark material filling embayments and irregularities in the margin of the other unit.

This lower unit differs markedly from the dark material, both in albedo and surface morphology. As can be seen from Fig.2 it has a ridged surface, each ridge being separated by deep channels. The general surface appearance is that of a once-viscous mass which flowed.

The southwesterly outcrop of the lower floor unit has a smoother surface and appears to have been subdued by a layer of fragmental material on its surface. The surface morphology of this unit resembles that of the floor units of large ray craters such as Tycho and Aristarchus and is considered to be of the same origin. In Tycho, the surface ridging on the lower floor unit clearly suggests that the unit was emplaced in a molten or semi-molten condition; large-scale polygonal fractures suggest contraction after emplacement, presumably by cooling; and concentric fractures near the unit's margin indicate some degree of compaction under gravity.

STROM and FIELDER (1968; see also Chapter 5) suggested that the crater floor of Tycho was flooded by viscous lava, and that the unit was volcanic in origin. A possible alternative to this may be considered: if Tsiolkovsky and other craters like it were of impact origin, layers of shock-melted ejecta may have accumulated, in particular, on the crater floors. These may have been formed of breccia at different stages of melting, analogous to the allochthonous fall-back breccias of the east Clearwater Lake crater in Canada (DENCE, 1968). Following this suggestion, the flow-like surfaces can be considered as recording viscous flow of the semi-molten breccia, and the fracturing from cooling of the material as a single unit. The

concentric fracture pattern may result from wholesale compaction-welding of this fragmental material.

The characteristic floor unit of Tsiolkovsky and similar craters can be shown to be younger than the major phase of fault-slumping of the inner crater walls, because the unit has a frontal scarp on the outer margin which enters embayments at the foot of the fault blocks. If the interpretation that the floor unit was formed, essentially, by fall-back of ejecta into the crater were correct, the implication would be that most of the internal slumping of the inner walls occurred before the bulk of the ejecta had fallen into the crater and solidified. Most of the terracing would thus belong to the crater-forming stage. Later landslipping would be essentially local. If the volcanic interpretation were correct, major slumping of the inner walls could be much younger.

Dark floor materials

This unit (Fig.1–4 and Fig.7) has a relatively smooth, level upper surface. Like other units associated with Tsiolkovsky it is pitted by small craters. Rilles are seen, especially near the edge where they tend to be concentric with the crater. Small mare ridges can just be identified on high-resolution pictures. The dark floor material is analogous in both location and form to the dark material of the nearside maria. Evidence from Lunar Orbiter pictures and, more recently, from returned samples, shows that this dark material is of igneous origin, consisting of basic lava. Deep fractures below the crater may well have aided the rise of magma and the onset of volcanism.

Structure

The rim, and parts of the ejecta blanket, are cut by a series of features interpreted as faults (Fig.9). This fault pattern also appears to have been adopted by the fault-slides of the inner, terraced walls. Two fault patterns may be recognised—one subradial to the crater and the other concentric with the lip. Faulting associated with the formation of large craters as well as the maria is a well established feature of the Moon (FIELDER, 1965). A strong fault pattern associated with Mare Imbrium has been mapped by several authors (see, for example, HACKMAN, 1966); and a similar pattern related to Tycho has more recently been described by SHOEMAKER et al. (1968).

The concentric pattern of the Tsiolkovsky rim and

inner walls probably consists of normal faults formed by partial resettling of the rim following its forcible uplift during crater formation. Although this pattern confirms broadly to a circular shape it presents a marked polygonal form reflected by the sides of the crater lip. The main directions followed by the sides of the polygon are approximately those of the "grid pattern" described by FIELDER (1963) and STROM (1964)—that is, northwest, northeast and north–south. These directions are roughly followed by the subradial fractures which cut both the ejecta blanket and older

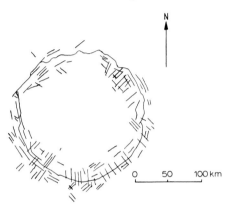

Fig. 9. Map showing the directions of principal fault structures on the rim of Tsiolkovsky.

crater rims near Tsiolkovsky. FIELDER and WARNER (1962) pointed out that, in general, subradial lineaments of lunar craters occur in bundles and that only the lineaments in the centre of the bundle strike towards the centre of the crater. This is the case for Tsiolkovsky. Examination of Fig.4 shows that there is also a tendency for crater chains and, indeed, the ejecta pattern itself, to follow the directions established for the fault patterns.

The close association of craters and maria with fault patterns, at least one of which shows evidence of movement over a protracted period of time (FIELDER and JORDAN, 1962; WILHELMS, 1968) has been used by a number of authors as evidence for an internal origin of maria and certain large craters, rather than for a catastrophic origin by impact. On the other hand, evidence from cratering experiments on both a large and small scale indicates that the apparent fault patterns associated with some lunar craters are consistent with a catastrophic origin. BAROSH (1969) has described explosion-produced fracture patterns in the Yucca Flat of the Nevada test site. These fractures not only form the expected concentric and radial patterns but are commonly aligned along certain preferential directions related to known faults, and are paralleled by sets of closely spaced joints in the bedrock. The frac-

ture patterns produced are not unlike those of Tsiolkovsky and other lunar craters. It may be demonstrated that much of the faulting in and around nuclear explosion craters results from the release of initial strain in the rocks. HAMILTON and HEALY (1969) showed that faulting was caused by release of natural tectonic strain at the Benham nuclear explosion, and that a sequence of earthquakes, lasting for some months after the explosion, was initiated. It is clear from this that the proposed catastrophic formation of Tsiolkovsky might be expected to initiate the intense faulting observed.

SHOEMAKER (1962) has pointed out that the trajectories of ejecta from high-energy explosion craters would be governed by structures in the bedrock at the explosion site. Laboratory experiments with high-velocity projectiles (GAULT et al., 1968) have adequately demonstrated that crater formation is a highly ordered process and that structures in the impacted medium can influence crater excavation. The final distribution of ejecta and secondary craters may thus tend to reflect structural trends in the bedrock. It would therefore appear that the structural evidence, although not excluding an internal origin for Tsiolkovsky, is consistent with previously cited evidence that it is of impact origin.

Tsiolkovsky has a prominent central peak. There is little direct evidence for the origin of this feature and it may be interpreted either as of internal or impact origin. In support of its being a volcano is the observed high density of craters in the summit platform (Fig.2). This excess of craters may indicate volcanism and suggest that the walls of part of the summit platform are the remains of an old caldera; but, in view of the very small area of the platform, crater counts must be treated with caution. The dark, mare-like crater floor has already been interpreted as representing volcanic activity within the crater; and the formation of the central peak could be related to this part of the crater's history.

Against this interpretation is the observation that central peaks are an important part of large, terrestrial impact craters (DENCE, 1968). Arguments (CHAPMAN and FIELDER, 1964) against an impact origin for central peaks have been based on the apparent tectonic control of these structures. The same arguments that were expressed previously about faulting outside the crater also apply here: uplift of the central peak would tend to follow a pre-existing stress field in the bedrock. The author considers it at least possible that the central peak of Tsiolkovsky is analogous to central peaks within known terrestrial impact craters; this would not preclude the possibility that volcanic edifices do occur in craters of this type, especially those which have a number of central eminences of different shape.

Small cratering and erosion

Present evidence suggests that the main erosive process on the Moon is meteoric bombardment. Accompanying this is the gradual, downhill creep of surface debris generated by impact breakdown of surface rocks. Assuming that the majority of small craters covering the lunar surface are impact craters, the number densities of small craters in individual units should be a measure of their relative ages. GUEST and MURRAY (1969) have given the number densities of craters for the main geologic units of Tsiolkovsky. Many of the counts were consistent with the stratigraphically established ages. However, variations of crater density were noted on individual units. Belts of higher density have been mapped (Fig.7) on the dark floor unit. The form of the craters and the nature of the belts suggested to Guest and Murray that they were caused by secondary impacts. Variations in number density on the rim unit fall mostly within the range of variation encountered on the dark floor unit. This would be expected if the variation in crater density were predominantly caused by an overprinting of secondary impact craters. Parts of the rim with anomalously low crater density are, without exception, areas of very hummocky terrain with high proportions of sloping ground. When shadows and overexposed areas are allowed for, the counts in these areas are still low and may be interpreted as being due to mass movement of surface material by creep.

The number density of craters in Tsiolkovsky is higher than that determined for Tycho by STROM and FIELDER (1968). Tycho has most of the major features of Tsiolkovsky, and the similarity between the craters shows them to have been produced by the same mechanism. Tsiolkovsky differs from Tycho in that its features have been softened more by erosion than in the case of Tycho. Craters older than Tsiolkovsky on the farside highlands—for example, crater A (Fig.6) which is cut by Tsiolkovsky—are, by comparison, more deeply eroded. POHN and OFFIELD (1969) have described a continuum of morphologic types of crater representing the degree of erosion suffered. This is found to

be consistent with age relationships known from stratigraphy. The conclusion that may be drawn from these observations is that many of the more denuded craters of the highlands once resembled Tsiolkovsky. It thus appears likely that most of them are of the same origin.

Conclusions

Systematic study of the geology of Tsiolkovsky leads to the conclusion that the present crater was formed by a single, high-energy explosion. The only natural phenomenon known at the present time that could have produced such an explosion is the impact of a large meteoroid, asteroid or comet. This conclusion agrees with that of many authors who have studied other craters of the same type in different parts of the Moon.

The following sequence of events may be envisaged for the formation of the different geologic units of Tsiolkovsky.

Crater forming stage

(1) Impact of a large body with the Moon.

(2) Crater excavation, spalling and overturning of lip to form part of raised rim.

(3) Deposition of ejecta blanket; formation of secondary craters from low angle ejected blocks; development of base surge.

(4) Collapse of outer rim to form ejecta flow (or slide).

(5) Partly accompanying (3) and (4) slumping of inner walls into the crater; fall-back of melted and shocked ejecta into crater, possibly to give a viscous allochthonous breccia unit on the crater floor; faulting following pre-existing basement trends.

(6) Continued fall-back of higher trajectory ejecta to produce secondary craters; continued movement on faults.

Post-crater forming stage

(7) Basic volcanism in crater floor to give the dark floor material and, possibly, the lower floor unit if this is volcanic rather than fall-back.

(8) Erosion by meteoric bombardment.

References

BALDWIN, R. B., 1963. *The Measure of the Moon*. University of Chicago Press, Chicago, Ill.

BAROSH, P. J., 1969. *Geol. Soc. Am., Mem.*, 110: 199.

CHAPMAN, R. G. and FIELDER, G., 1964. *Observatory*, 84: 23.

DENCE, M. R., 1968. In: B. M. FRENCH and N. M. SHORT (Editors), *Shock Metamorphism of Natural Materials*. Mono Book, Baltimore, Md.

FIELDER, G., 1963. *Quart. J. Geol. Soc. London*, 119: 65.

FIELDER, G., 1965. *Lunar Geology*. Lutterworth, London.

FIELDER, G. and JORDAN, C., 1962. *Planetary Space Sci.*, 9: 3.

FIELDER, G. and WARNER, B., 1962. *Planetary Space Sci.*, 9: 11.

GAULT, D. E., QUAIDE, W. L. and OBERBECK, V. R., 1968. In: B. M. FRENCH and N. M. SHORT (Editors), *Shock Metamorphism in Natural Materials*. Mono Book, Baltimore, Md.

GUEST, J. E. and MURRAY, J. B., 1969. *Planetary Space Sci.*, 17: 121.

HACKMAN, R. J., 1966. *U.S., Geol. Surv., Map I-463*.

HAMILTON, R. M. and HEALY, J. M., 1969. *Bull. Seismol. Soc. Am.*, 59: 2271.

MURRAY, J. B. and GUEST, J. E., 1970. *Mod. Geol.*, 1: 149.

OFFIELD, T. W. and POHN, H. A., 1969. *U.S., Geol. Surv., Rept. Astrogeol.*, 13.

POHN, H. A. and OFFIELD, T. W., 1969. *U.S., Geol. Surv., Rept. Astrogeol.*, 13.

SHOEMAKER, E. M., 1962. In: Z. KOPAL (Editor), *Physics and Astronomy of the Moon*. Academic Press, London, p.283.

SHOEMAKER, E. M., BATSON, R. M., HOLT, H. E., MORRIS, E. L., RENNILSON, J. J. and WHITAKER, E. H., 1968. *NASA Rept.*, SP-173.

SHREVE, R. L., 1966. *Science*, 154: 1639.

STROM, R. G., 1964. *Commun. Lunar Planetary Lab. Univ. Arizona*, 2: 205.

STROM, R. G. and FIELDER, G., 1968. *Nature*, 217: 611.

WILHELMS, D. E., 1968. *U.S., Geol. Surv., Map I-548*.

7

Origins of lunar craters

R. J. FRYER

Introduction

It is now generally agreed that both impact and volcanic processes have helped to shape the lunar surface. It remains to establish their relative importance as a function of scale of feature and position on the Moon.

This requires skills derived from a number of disciplines. If it were possible to determine unambiguously the extent, and nature, of lunar volcanism it should also be possible to draw far-reaching conclusions about the internal thermal history of the Moon and, by implication, of the other terrestrial planets. This being the case, it is profitable to consider the types of formation that are to be expected on the Moon and which can be loosely classified as "craters". Table I lists the cratering processes, and their major characteristics, that may be assumed to act, or to have acted, on the lunar surface. Clearly the range of possible morphology is large and features not usually classed as craters—for example, calderas—are included because of their superficial similarity to true craters.

Numerous criteria can be applied to distinguish between crater types, but the criteria chosen must depend on the type of photographic coverage available. Four kinds of test are described below.

TABLE I

POSTULATED LUNAR CRATERING MECHANISMS

Mechanism	Cause	Typical characteristics when fresh
Single "explosion"	meteoric primary impact; high-velocity secondary impact; volcanic explosion	radial symmetry; rays; secondary craters; base surge features around larger craters; well developed rim; tectonic control of shape; terracing (?)
Multiple "explosion"	multiple primary meteoric impact; multiple high-velocity secondary impact; volcanic processes (e.g., maar; explosive caldera)	rays; secondary craters; base surge features around larger craters; well developed rim; tectonic control of volcanic features; terracing
Low-velocity excavation	low-velocity secondary impacts; tertiary impacts	bilateral symmetry; ray plumes; chains or elongated clusters of craters
Collapse	quiescent caldera subsidence	ignimbrite/lava flows; flat floors (?); tectonic control of shape; large scale (> 1 km)
	rubble drainage into cavities and fissures	conical or convex-up profile; possible bilateral symmetry; chains of depressions; tectonic control of chain directions
	collapse of lava surface into tubes	irregular depressions; chains of depressions; chains controlled by flow pattern of fluid lava; surrounding fissure system; small size (< 1 km)
Emission of material	cinder cones	cratered peak; topographic high; tectonic control of shape and position; lava flows (?)
	shield volcanoes	cratered peak; broad topographic high; lava flows; tectonic control of shape and position
	ring dykes	lava flows; incomplete rim (?); flat floors (?)

Tests to distinguish between crater types

Distributional methods

Craters are not distributed at random on the lunar surface, but show great differences in number density between marial and highland regions. The maria themselves are not randomly distributed, but are concentrated on the Earth-side hemisphere. Among others, HARTMANN (1967) has shown that, on log–log plots, separate crater diameter versus number-density curves for each type of terrain are approximately parallel, the number density in the highlands generally exceeding that in the maria by more than an order of magnitude.

Such a non-uniformity reveals more about the relative youth of the marial surfaces than about the craters themselves, however, and, if meaningful conclusions are to be drawn about the craters, it is neces-

TABLE II

ESTIMATED PROPORTION *(P)* OF TECTONICALLY ALIGNED ENDOGENIC CRATERS IN THE DIFFERENT AREAS OF THE LUNAR MARIA[1]

Mare	Crater number density[2]	Minimum crater diameter mapped (m)	P_{min} (%)	P_{max} (%)
Crisium	15	670	45	58
Fecunditatis	22	460	32	63
Frigoris	7	405	29	43
Frigoris	33	703	43	96
Humorum	11	320	21	91
Humorum	9	335	61	86
Imbrium	10	700	44	55
Imbrium	11	700	29	37
Medii	23	116	24*	33*
Medii	23	116	0**	6**
Nubium	28	685	48	67
Nubium	15	663	26	62
Orientale	30	650	39	48
Procellarum	17	130	17	79
Serenitatis	12	643	49	57
Serenitatis	10	633	25	32
Serenitatis	10	317	16	22
Serenitatis	14	643	48	59
Serenitatis	8	634	41	66
Serenitatis	8	317	76	100
Smythii	5	360	24	31
Tranquillitatis	8	100	52	66
Tsiolkovsky	11	380	31	39

[1] After FIELDER et al. (1970).
[2] Adopted total number of craters $\geq$ 700 m in diameter per 1,000 km² of surface.
* Eumorphic ("sharp") craters only.
** Submorphic ("soft") craters only.

sary to investigate them on the different types of terrain separately. When this is done it is obvious that the maria themselves do not form a homogeneous unit, since crater number densities (Table II) vary between—and even within— individual maria. Even more subtle trends are detectable. For instance, FIELDER (1965) used Earth-based photographs to show that, when equivalent terrain types were considered, there was a systematically higher number density of craters in the "following" (eastern) half as compared with that half of the visible disk that "leads" in its orbit around the Earth.

Three main difficulties of counting craters from a photograph can be enumerated: incompleteness of observation; ambiguities in diameter measurements; and obliteration of smaller craters by younger and larger ones. The first two are critically dependent on crater morphology. Submorphic craters (Chapter 1) are difficult to detect under some lighting conditions, and predominate among those "lost" at sizes near the photographic limit of resolution. It is also difficult to assign diameters to such craters, since there is no sharp boundary between their interiors and the surrounding terrain. This is rarely important for positional studies, but becomes so when size distributions are considered. Incompleteness of observation will be accentuated in high-relief terrain where lighting conditions vary from point to point. The obliteration of smaller craters by larger ones is important only at high crater densities, such as occur in the highlands.

Highland distributions

Obliteration is a difficult process to treat analytically, and no firm results based on the positional distribution of all highland craters have so far been achieved. ARTHUR (1954) demonstrated that, for high crater densities, the positional distribution of remaining craters differed widely from a Poisson distribution, even for completely random impacts. Progressive erosive forces can so alter the form of a crater that it becomes unrecognisable without being totally obliterated. For these reasons, investigations of the distribution of all craters in the highlands must at present be interpreted with caution.

An ingenious partial avoidance of these problems has been reported by TITULAER (1969). In his investigation he considered "parasite" (overlapping) and "host" (overlapped) craters. Using the catalogue of ARTHUR et al. (1963, 1964, 1965, 1966) he separated out all host craters larger than 30 km in diameter which lay within 70° of the sub-Earth point and, for each crater, determined the corresponding number of parasites

Fig.1. The near-side distribution of "host" craters, larger than 30 km in diameter, which have at least twice the corresponding average number of "parasite" craters larger than 3 km in diameter. (After TITULAER, 1969.)

larger than 3 km in diameter. From the list of hosts so formed he then segregated those having more than twice the mean number of parasites appropriate to their size. These host craters, located in Fig.1, are almost entirely confined to the northwestern part of the southeast highlands. This concentration cannot be ascribed to limb losses of small craters since some examples of prolific hosts are detectable out to the 70° limit. Further work by FRYER and TITULAER (1970), using Orbiter 4 photographs, has shown that the southeast boundary of this concentration can also be detected

as a change in the number densities of craters in the size range 350–2,500 m. Flat crater floors and intercrater areas were investigated separately, and both showed the change.

Mare distributions

The distributions of craters within the maria are more susceptible to direct investigation. Number densities are, in general, so low that obliteration may be neglected (FIELDER et al., 1970). Hence a reasonable working hypothesis is that primary craters have a

Poisson distribution in any given mare unit. This primary distribution will be complicated by the addition of distributions arising from any of the other processes listed in Table I. Apart from impact, most crater-forming processes will not generate random distributions. Lava flows, destruction by impact, or selective shakedown of rubble rims by seismic activity will all change the combined distribution, the effects usually being greatest for the smallest craters.

To deal with such a combination of processes by a statistical method it is necessary to make a number of assumptions. Those adopted by FIELDER et al. (1970) are typical: (*1*) primary impacts have a Poisson distribution in the area considered; (*2*) obliteration is negligible; (*3*) crater alignments with parallel trends over large areas are of endogenic, rather than of secondary impact, origin; (*4*) crater alignments associated with recognisable secondary impact patterns are not necessarily of internal origin; and (*5*) loss of small craters during counting is proportional to the local number density of craters.

The application of these assumptions may be illustrated by an examination of the Flamsteed region

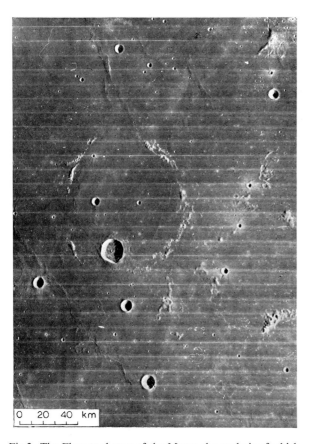

Fig.2. The Flamsteed area of the Moon, the analysis of which, in terms of crater position distribution, is discussed in the text. (Orbiter 4 photo, NASA.)

of Oceanus Procellarum, one of the areas considered by FIELDER et al. (1970) in a general investigation of marial areas. The area covered by Fig.2 was selected and all craters with photographic diameters of 0.5 mm (130 m true diameter) or greater were mapped (Fig.3, 4A). A comparison map was made by randomly distributing craters (Fig.4B) which had the same diameter distribution; however, the seven largest craters were placed in the same positions as in the original field, to aid interpretation.

In order to investigate non-random clustering over the entire region, the area was divided into 496 equal areas, each containing about 60 craters, and the numbers of crater centres falling in each area were compared with those expected from a Poisson process. That crater clustering is very important in this region may be demonstrated in a number of ways. Fig.3 includes contours corresponding to twice the mean number density of craters. Application of the chi-squared test to the 496 area counts indicated that the probability of the observed distribution being drawn from a Poisson distribution was negligible. This is illustrated in Fig.5, which shows a comparison of the observed distribution of the numbers of crater centres per area with the predicted distribution, and also with an analogous distribution derived from a count of the randomly redistributed craters. An alternative way of revealing clustering is illustrated in Fig.4. The clustering was determined by measuring the light transmitted through elements of a negative copy of the map and, separately, its random simulation. The clusters thus correspond to regions in which a large proportion of the surface is covered by craters. One may ignore "clusters" known to be generated by large craters.

With assumptions (*1*) and (*5*), it appears that the area contains craters of other than primary impact origin since the departure from randomness cannot be ascribed to localised crater losses. The dense clusters shown in Fig.3 appear to be genetically unrelated to large craters nearby; neither can they be identified with prominent ray systems. The implication is, therefore, that they are not clusters of secondaries, but are volcanic in origin, occupy areas of poor seismic coupling to the surroundings, or are older areas that have escaped obliteration by flooding.

Assumption (*3*) provides a means of estimating the amount of volcanicity that an area has supported. To search for crater alignments a statistical sampling technique was adopted by FIELDER and MARCUS (1967) and was used also by FRYER (1968) and FIELDER et al. (1970). All these authors used long, narrow sample areas (strips), the strips in each set being parallel, and

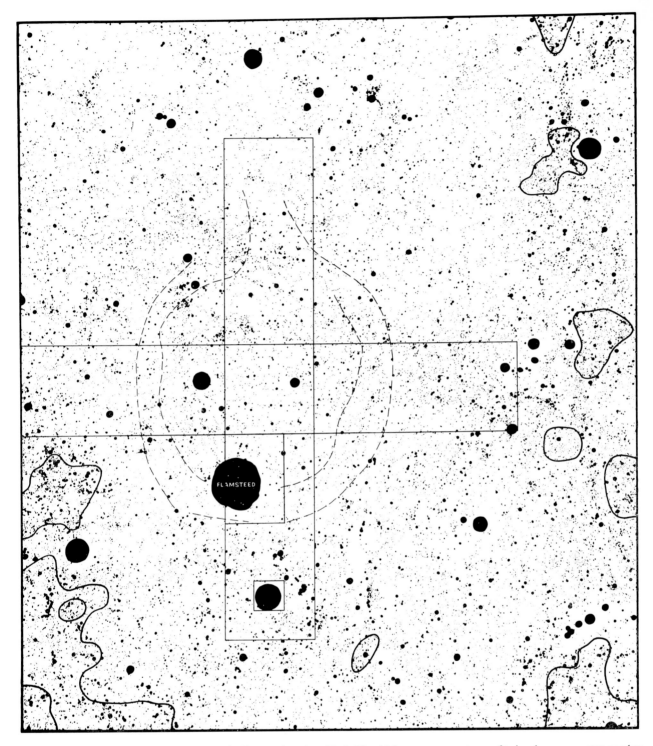

Fig.3. A map of all craters larger than 130 m in diameter based on Fig.2. The thick curves are contours of twice the mean crater number density. The rectangles delimit the areas used in obtaining the data plotted in Fig.8. (After FIELDER et al., 1970.)

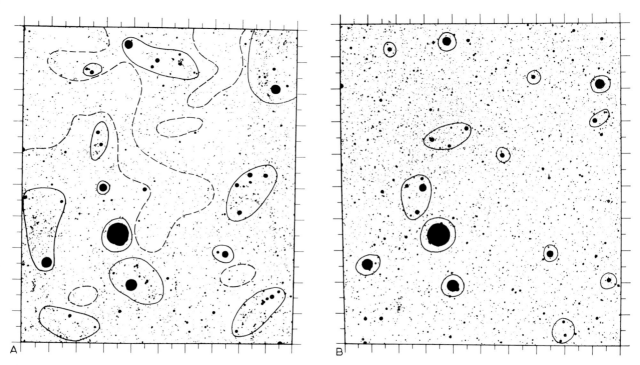

Fig.4. Photometrically derived contours of equal fractional coverage of the lunar surface by craters. A. Original version. B. Simulated version. Solid lines correspond to twice the mean photometric density, dashed contours to half the mean. (After FIELDER et al., 1970.)

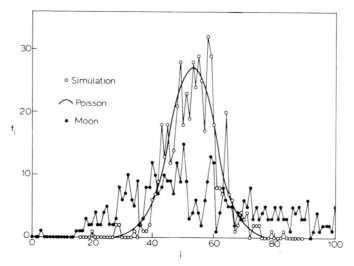

Fig.5. Comparison of square counts of the original (filled circles) and simulated (open circles) versions of the map of craters (Fig.3) with the theoretically predicted Poisson curve. The tail of the distribution curve of the original version is detectable beyond $i = 160$, where i is the number of craters per element area. The ordinate f_i is the number of element areas containing i craters. (After FIELDER et al., 1970.)

employed a number of sets to cover the full range of azimuths.

The "index of dispersion", V, was then computed from:

$$V = S_{N^2}/\overline{N}$$

where:

$$\overline{N} = \sum_{i=1}^{r} N_i/r$$

and:

$$S_{N^2} = \sum_{i=1}^{r} (N_i - \overline{N})^2/(r - 1)$$

$N_1 \ldots N_r$ being the number of crater centres counted in each of r parallel strips.

When V is plotted as a function of azimuth, θ, preferential crater alignments manifest themselves as peaks, generalised clustering without preferential alignments as a raising of the curve above the expected value of $V = 1$, and exceptional uniformity of distribution as a depression of the curve below $V = 1$. Chains which are not parallel over the whole area do not give rise to significant peaks. Such a test has the advantage of being readily applicable to an area containing many points, and of being able to pick out chains superimposed on a background. Visual tests for chains of craters—such as those FIELDER (1968) applied to Ranger photographs—are useful in the case of areas with few points.

FIELDER et al. (1970) compared their observed V values with the V values pertaining to computer generated random models having known proportions of chained craters, the size and shape of the sample areas being equivalent to those on the original maps. In this way an estimate was made of the proportions of internal craters in the various areas investigated. It was estimated that, for the whole area described by Fig.2, between 48 and 80% of the craters were of internal origin.

The map of Fig.3 was split into six sections, as shown in Fig.6, and each area was investigated separately. The results are shown in Fig.7 and Table III.

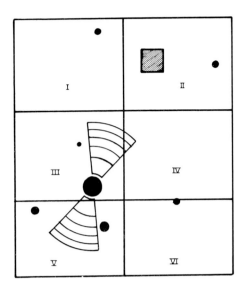

Fig.6. The six sections of Fig.3 discussed in the text. The annular segments were counted for the preparation of Fig.9. The stippled area shows the sample size used in the preparation of the photometric contours (Fig.4).

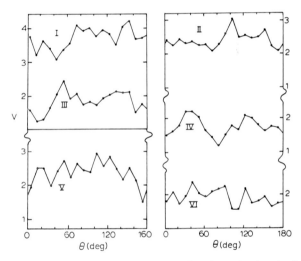

Fig.7. "Index of dispersion", V, as a function of azimuth, θ, for the six sections (Fig.6) of Fig.3. (After FIELDER et al., 1970.)

TABLE III

ESTIMATED PROPORTIONS (P) OF TECTONICALLY ALIGNED
ENDOGENIC CRATERS IN THE SIX SECTIONS OF FIG.6*

Section	Area (km^2)	P_{min} $(\%)$	P_{max} $(\%)$
I	12,250	30	79
II	12,550	20	33
III	12,450	27	31
IV	12,450	17	21
V	10,500	25	37
VI	10,500	19	30

* After FIELDER et al. (1970).

Clearly, the area of Fig.2 is by no means homogeneous in terms of crater density. The whole area contains clusters of craters, though clustering is particularly strong in section I, which also has the highest uncertainty in the estimated proportion of endogenic craters.

The lowest proportion of internal craters occurs in section IV which also shows chaining in its V versus θ plot. It seems that chains dominate the type of clustering in this section in contrast to section I, where amorphous clusters predominate.

Chaining directions vary from section to section, and are commonly related to the local trends of wrinkle ridge segments. Application of the V test to a small

portion of the interior of Flamsteed P by FRYER and TITULAER (1969) gave no sign of chaining, but indicated that craters in the diameter range 13–380 m were distributed more uniformly than expected on a random model. They regarded this tendency to uniformity as evidence for collapse depressions in a lava surface. This would imply that the interior of Flamsteed P was younger than the exterior. Counts (Fig.8) along the strips indicated in Fig.3 confirm this impression, since the number density generally changes sharply at the wall of Flamsteed P.

There is no analogous distribution of small secondary craters in the case of Flamsteed P. Counts of the number densities of craters in the annular segments

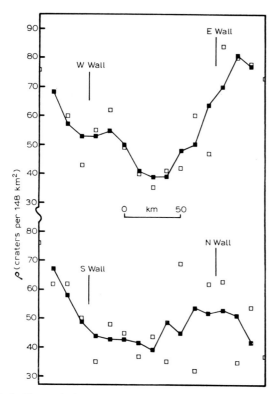

Fig.8. The variation of number density of craters across the Flamsteed P ring obtained by counts in the regions shown in Fig.3. (After FIELDER et al., 1970.)

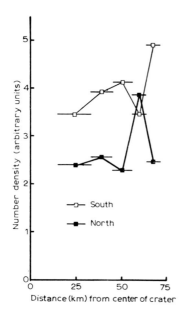

Fig.9. The variation of number density of craters with distance from Flamsteed, obtained by counts in the annular areas shown in Fig.6.

shown in Fig.6 and plotted in Fig.9 show no peaks such as might be expected for secondary craters (see, for example, SHOEMAKER, 1962). If secondaries were ever present they could have been obliterated by lava flows.

In summary, the area covered by Fig.2 appears to have supported widespread endogenic cratering and to have been subjected to localised obliteration processes, probably impacts. There were probably flows of lava with numerous collapse depressions.

Similar investigations of other mare regions by FIELDER et al. (1970) produced the data included in Table II. It is seen from Table II that the estimated amount of internal cratering varies greatly between the various maria.

Geologic–morphologic methods

Techniques standard in photogeology (MILLER and MILLER, 1961; ALLUM, 1970) have been adapted to the study of the Moon, notably by the U.S. Geological Survey in their series of lunar geological charts. The Lunar Orbiter missions in the 1960's have produced an immense amount of photographic data, which are being used both in the revision of Earth-based lunar geologic maps and in the detailed investigation of individual features. A special study of the large far-side crater Tsiolkovsky has been presented in Chapter 6 where the structure has been interpreted as basically of impact origin. In an independent geologic assessment of Tsiolkovsky, EHRLICH et al. (1970) refute the impact origin. Examining the similar structures Tycho and Aristarchus, STROM and FIELDER (1968; see also Chapter 5) concluded that at least parts of the respective formation processes were characterised by widespread volcanic eruptions.

In the morphologic method of KUIPER (1965) and KUIPER et al. (1966) different types of crater (impact, collapse, and so on) are differentiated on grounds of shape.

Statistical–morphologic methods

The studies of Tsiolkovsky and Tycho mentioned above drew upon high-resolution photographic coverage which enabled detailed studies of relatively small-scale structures to be made. However, the vast majority of craters visible on photographs are small and are not, therefore, amenable to detailed individual examination. Nevertheless, following BALDWIN (1963), it is possible to perform statistical studies of dimensions characteristic of such craters.

Baldwin investigated the dependence of depth, apparent rim width and rim height on crater diameter for lunar craters, and attempted to link the deduced relations with the corresponding relations for terrestrial craters of known origin. The subsequent emphasis on the log–log plot of depth versus diameter is unfortunate in that it is not a straight line but is composed essentially of two individual portions—one for terrestrial explosion craters and the other representing large lunar craters, covering different diameter ranges. It is true that the sections are linked by a small number of terrestrial impact structures but, bearing in mind the uncertainties in the original dimensions—particularly the depth— of these structures, the data are not convincing. Another criticism of Baldwin's interpretation has been made, among others, by GREEN and POLDERVAART (1960) and STEINBERG (1966); namely, that some terrestrial calderas also lie on the curve.

Baldwin's second correlation, a log–log plot of apparent rim width versus diameter, is more satisfactory. Such a plot (fig.5 in Chapter 6) is seen to be closely linear, probably as a result of impact cratering mechanics (see, for example, GAULT et al., 1968). Calderas and terrestrial volcanic explosion craters associated with pyroclastic flows do not lie on the same line, though terrestrial hypervelocity impact or explosion craters do so. Whilst measurements of this sort cannot be used to settle the origin of any particular feature, indications of the distribution of likely origins among groups of lunar features can easily be gained. Methods of this type have the advantage that only plan views on single photographs are required.

Other readily measured parameters may be considered. The work by CHAPMAN and FIELDER (1964), on the displacements of "central" peaks from the geometric centres of large craters, may be worthy of further investigation. These authors detected preferential displacements approximately along the directions of the grid system. The work of GAULT et al. (1968) on scaled, hypervelocity impact craters in inhomogeneous media has demonstrated that regional structure can strongly influence the rim form of an impact crater. Whether tectonic control of a non-volcanic "central" peak is probable has not been demonstrated.

Single measurements of typical parameters do not take advantage of all the potential information. The shapes of the lips of craters have been measured by several authors. Morphologic comparisons of craters can be made sensitive to different aspects of shape—for instance, to over-all ellipticity or to the degree of crenellation of the rim.

Such tests have been used (FIELDER, 1961; FIELDER and JORDAN, 1962; RONCA and SALISBURY, 1966; ADLER and SALISBURY, 1969) to investigate the stress history of the lunar surface on the assumption that craters were originally circular. Other investigations of crater circularities, bearing more directly upon the

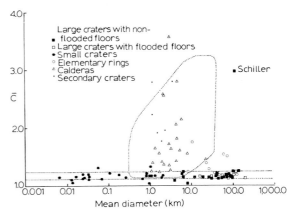

Fig.10. The "circularity", $\bar{C}$, of a variety of lunar formations plotted as a function of mean diameter. The parallel, dotted lines indicate the analogous standard deviation limits in $\bar{C}$ for a sample of terrestrial meteorite and artificial explosion craters. The dotted outline shows the approximate limits of a sample of terrestrial calderas and collapse pits. The individual symbols indicate the assigned class of each formation based on independent criteria. (After MURRAY and GUEST, 1970.)

origin of the craters themselves, have been reported in Chapter 6. A survey of the circularities of lunar and terrestrial craters has been reported by MURRAY and GUEST (1970; see Fig.10). Using rim shape alone, no diagnosis may be made of the origin of any individual feature. On the other hand, it has been demonstrated that lunar formations originated in a variety of ways. Murray and Guest attempted to identify the origin of individual features by morphologic criteria other than circularity, and these identifications are also included in Fig.10. The identifications are consistent with the observed distribution of circularities.

Statistical–size methods

When any set of craters is mapped it is possible to obtain some measure of the distribution of sizes within the set. The data so obtained are subject to the difficulties arising from incompleteness and diameter measurement discussed earlier in this chapter. In addition, grouping of craters into diameter intervals can introduce complexities into subsequent statistical analyses. Subjective effects are very important in counts made on photographs. FIELDER et al. (1970) demonstrated that, except under carefully controlled conditions, measurements of the same area by different workers may differ widely, even in areas where the number density of craters is relatively low. The problems inherent in making meaningful comparisons between observed and theoretical diameter distributions, or intercomparisons between published data, are therefore great.

In the highland regions such difficulties are particularly important, since obliteration effects complicate the theory still further. An attempt has been made by MARCUS (1966a–d) to account statistically for changes in densely cratered areas; but mapping problems have so far prevented any final test of his results. Earlier work (MARCUS, 1964) on the diameter distribution of "clean" (undamaged) craters can be applied more readily, though mapping difficulties are still serious.

For marial areas it is customary to present diameter data as log–log plots of cumulative number density versus diameter. Such plots are often, but not invariably, found to be linear. Large departures from linearity, provided the sample is large enough, may be significant. DOLLFUS et al. (1970), for instance, interpreted counts made on Mariners 6 and 7 Mars photographs (when taken in conjunction with certain morphologic considerations) as demonstrating a former, dense Martian atmosphere. There are many pitfalls in such methods and, for most purposes, the use of incremental counts is to be preferred (CHAPMAN and HAEFNER, 1967). Such counts must then be interpreted in conjunction with other data.

Conclusion

Evidence exists for the presence on the Moon of craters formed by all the processes listed in Table I. Each of the analytical techniques outlined above is capable of illuminating some aspect of the problem; taken together they can be powerful grounds for the acceptance or rejection of a hypothesised process. The use of multiple techniques for investigating the lunar surface is now becoming widely accepted and the results from the application of such techniques can have an important bearing on our understanding of large-scale internal processes and even on discussions of the thermal history of the Moon.

References

ADLER, J. E. M. and SALISBURY, J. W., 1969. *Icarus*, 10: 37.

ALLUM, J. A. E., 1970. *J. Brit. Interplanetary Soc.*, 23: 297.

ARTHUR, D. W. G., 1954. *J. Brit. Astron. Assoc.*, 64: 127.

ARTHUR, D. W. G., AGNIERAY, A. P., HORVATH, R. A., WOOD, C. A. and CHAPMAN, C. R., 1963. *Commun. Lunar Planetary Lab. Univ. Arizona*, 2: 71.

ARTHUR, D. W. G., AGNIERAY, A. P., HORVATH, R. A., WOOD, C. A. and CHAPMAN, C. R., 1964. *Commun. Lunar Planetary Lab. Univ. Arizona*, 3: 1.

ARTHUR, D. W. G., AGNIERAY, A. P., PELLICORI, R. H., WOOD, C. A. and WELLER, T., 1965. *Commun. Lunar Planetary Lab. Univ. Arizona*, 3: 61.

ARTHUR, D. W. G., PELLICORI, R. H. and WOOD, C. A., 1966. *Commun. Lunar Planetary Lab. Univ. Arizona*, 5: 1.

BALDWIN, R. B., 1963. *The Measure of the Moon*. University of Chicago Press, Chicago, Ill.

CHAPMAN, C. R. and HAEFNER, R. R., 1967. *J. Geophys. Res.*, 72: 549.

CHAPMAN, R. G. and FIELDER, G., 1964. *Observatory*, 84: 23.

DOLLFUS, A., FRYER, R. J. and TITULAER, C., 1970. *Compt. Rend.*, 270: 424.

ERLICH, E. N., GORSHKOV, G. S., MELEKESTSEV, I. V. and STEINBERG, G. S., 1970. *Mod. Geol.*, 1: 197.

FIELDER, G., 1961. *Planetary Space Sci.*, 8: 1.

FIELDER, G., 1965. *Monthly Notices Roy. Astron. Soc.*, 129: 351.

FIELDER, G., 1968. *Planetary Space Sci.*, 16: 501.

FIELDER, G. and JORDAN, C., 1962. *Planetary Space Sci.*, 9: 3.

FIELDER, G. and MARCUS, A., 1967. *Monthly Notices Roy. Astron. Soc.*, 136: 1.

FIELDER, G., FRYER, R. J., TITULAER, C., HERRING, A. K. and WISE, B., 1970. *Phil. Trans. Roy. Soc.* (In press.)

FRYER, R. J., 1968. *Planetary Space Sci.*, 16: 805.

FRYER, R. J. and TITULAER, C., 1969. *Commun. Lunar Planetary Lab. Univ. Arizona*, 8: 51.

FRYER, R. J. and TITULAER, C., 1970. *Earth Planetary Sci. Letters*, 9: 6.

GAULT, D. E., QUAIDE, W. L. and OBERBECK, V. R., 1968. In: B. M. FRENCH and N. M. SHORT (Editors), *Shock Metamorphism of Natural Materials*. Mono Book, Baltimore, Md.

GREEN, J. and POLDERVAART, A., 1960. *Intern. Geol. Congr., 21st, Copenhagen, 1960*, 21: 15.

HARTMANN, W. K., 1967. *Commun. Lunar Planetary Lab. Univ. Arizona*, 6: 39.

KUIPER, G. P., 1965. *Jet Prop. Lab., Tech. Rept.*, 32-700 (2): 9.

KUIPER, G. P., STROM, R. G. and LE POOLE, R. S., 1966. *Jet Prop. Lab., Tech. Rept.*, 32-800(2): 35.

MARCUS, A. H., 1964. *Icarus*, 3: 460.

MARCUS, A. H., 1966a. *Icarus*, 5: 165.

MARCUS, A. H., 1966b. *Icarus*, 5: 178.

MARCUS, A. H., 1966c. *Icarus*, 5: 190.

MARCUS, A. H., 1966d. *Icarus*, 5: 590.

MILLER V. C. and MILLER C. F., 1961. *Photogeology*. McGraw-Hill New York, N.Y.

MURRAY, J. B. and GUEST, J. E., 1970. *Mod. Geol.*, 1: 149.

RONCA, L. B. and SALISBURY, J. W., 1966. *Icarus*, 5: 130.

SHOEMAKER, E. M., 1962. In: Z. KOPAL (Editor), *Physics and Astronomy of the Moon*. Academic Press, London.

STEINBERG, G. S., 1966. *Soviet Phys., Dokl. (English Transl.)*, 10: 1006.

STROM, R. G. and FIELDER, G., 1968. *Nature*, 217: 611.

TITULAER, C., 1969. *Commun. Lunar Planetary Lab. Univ. Arizona*, 8: 63.

8

Photometric studies

L. WILSON

Introduction

Study of the photometry of an astronomical body involves the measurement of the light scattering properties of the surface of the body. The albedo of a surface measures its light scattering efficiency and is defined here as the apparent brightness of an illuminated surface element divided by the apparent brightness of a perfectly diffusing screen placed in the same orientation and at the same position as that element. Curves of the albedo of a planetary object as a function of wavelength may be used in attempts to identify the minerals present on its surface (SINTON, 1967).

The Moon is the only planetary body without an atmosphere that has been well observed photometrically: the present account deals with the methods of determining the light scattering properties of the lunar surface, and the interpretation of its scattering functions. The main features of the light scattering laws pertaining to the lunar surface were deduced before 1930 as a result of the work of WIRTZ (1915), ROSENBERG (1921), BARABASHEV (1922), MARKOV (1924) and ÖPIK (1924). It became clear that: (1) lunar surface elements had a generally low normal albedo (in the range 0.05–0.18) and (2) most of the sunlight incident on the Moon was scattered back towards the Sun.

The photometric properties of the Moon can be expressed most readily in terms of the angles shown in Fig.1. The angles of incidence, ι, and emergence, ε, of light at a point on the lunar surface are, respectively, the angles between the surface normal at that point and the lines joining the point to the source of illumination and to the observer. The angle of vision, ϕ, is the angle subtended at the point by the centre of the Sun and the observer; the azimuthal angle, z, is the

angle, in the plane tangent to the surface, between the planes[1] of incidence and emergence. The above conclusion (2) can now be re-stated: the apparent brightness of all surface features is a maximum when $\phi = 0°$.

The most important sets of photometric data for examining the physical properties of selected lunar regions are the catalogues, listing brightness as a function of ϕ, of ÖPIK (1924), BENNETT (1938), SHARANOV (1939), MARKOV (1950), FEDORETZ (1952) and VAN DIGGELEN (1958), and the isophote maps of SAARI and SHORTHILL (1967).

As a consequence of the Moon's synchronous rotation and the lunar libration, the Earth appears to move by only about 8° in the lunar sky, so that there is a variation of no more than this amount in the angle ε for any one location on the Moon. Hence, to determine the complete scattering properties of lunar material as a function of ι, ε and ϕ, it is necessary to combine observations of different lunar regions having the same type of terrain; the different regions may have somewhat different albedoes, however.

This procedure was adopted by ORLOVA (1956), who used the catalogue of Fedoretz. However, HAMEEN-

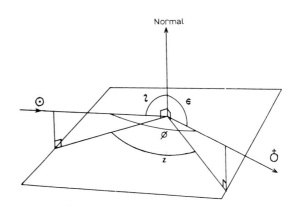

Fig.1. The relation between the angles ι, ε, ϕ and z in the general case.

---

[1] These planes are formed, respectively, by the incident (or emergent) rays and their normal projections on the plane tangent to the surface at the point in question.

ANTTILA (1967) noticed that the functions derived by Orlova do not satisfy the symmetry requirement of MINNAERT (1941). This symmetry law can be stated: "the brightness, $B(\phi, \iota, \varepsilon)$ of any point on the lunar surface has the property that the value of the function $B(\phi, \iota, \varepsilon) \cos \varepsilon$ is unchanged if ι and ε are interchanged". When this test is applied to Orlova's tabulated functions it is seen that there is an asymmetry which varies systematically with ϕ (WILSON, 1968). This implies some anomaly in Orlova's normalisation of Fedoretz's brightness data. For this reason, the complete photometric functions have been derived here for the two main types of lunar terrain—mare and highland—from the observational data of Öpik, Bennett and Fedoretz (WILSON, 1968). Each of the three observers provided large numbers of brightness measurements. Their data are now analysed separately: a comparison of the resulting functions of brightness versus ι, ε enables any systematic differences between the different observers' photometric scales to be detected.

Reduction of data

A series of small lunar areas lying near the brightness equator were selected as being typical (or nearly so) of mare material, on the one hand, and continental material, on the other. The curves of brightness as a function of ι were plotted from the named catalogues— a total of 920 measurements— and normalised to have the same brightness at $\phi = 0°$. In fact, the brightness values at $\phi = 0°$ are not known because $\phi = 0°$ corresponds to a total lunar eclipse. However, it can be shown that MINNAERT's (1941) symmetry rule leads to the requirements that, in the above notation:

$$B(\phi = 0, \iota, \varepsilon) = \text{constant for all } \iota, \varepsilon \qquad (1)$$

Thus, one is justified in making this normalisation. The calculation has to be carried out using the observed brightness values near to $\phi = 0°$ rather than at $\phi = 0°$; but it is found, retrospectively, that this approximation leads to a very small error in the final photometric functions. When all the observed brightness curves have been normalised, values of brightness can be read off at uniform intervals of ι and plotted on a new family of graphs, one graph for each chosen value of ι. Smooth curves are drawn through the plotted points, and brightness values read off at uniform intervals in ε. In this way a convenient tabulation of the scattering functions with $10°$ intervals of ι and $5°$ intervals of ε can be generated. Such tables have

been compiled for mare and highland areas, taken separately, from each of the catalogues of ÖPIK (1924), BENNETT (1938) and FEDORETZ (1952).

Determination of relative errors in brightness

The presence of systematic errors in photographic photometric measurements is commonplace. The relative values of the systematic errors in brightness of any pair of observers can be determined by relating their measured brightnesses of the same type of lunar terrain for the same pair of angles (ι, ε). Fig.2 and Fig.3 show the respective correlation diagrams for measurements on mare areas by Fedoretz and Bennett, and Öpik and Bennett. The plot for Fedoretz and Bennett is a straight line, indicating that the photometric scales of these observers are linear (or both non-linear in the same way, which is very unlikely); the line relating the data of Öpik to those of Bennett is a curve, suggesting that there is some non-linear error in the data of Öpik—a possibility that ÖPIK (1924) himself suggested in his original paper. The existence of a finite intercept in Fig.2 shows that the data of either Fedoretz or Bennett, or both, are subject to constant additive error. Corresponding graphs using data from continental regions have been prepared and confirm these conclusions.

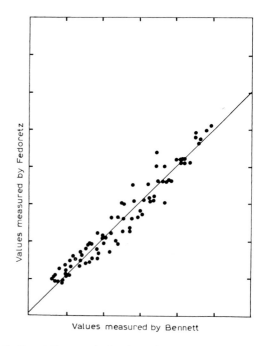

Fig.2. Correlation graph showing values obtained by FEDORETZ (1952) and BENNETT (1938) for the average brightness of mare areas under various conditions of illumination and observation. The scales are proportional to the original measurements.

The errors in brightness are characteristic of all the major catalogues of lunar photometric measurements. VAN DIGGELEN (1958) compared the data obtained by two or more observers for certain crater floors and derived correlation graphs, like Fig.2 and Fig.3, relating the scales of other workers to his own. Although such graphs can be used to combine all available measurements to form any one scale, it is not possible to decide which, if any, is free from systematic error.

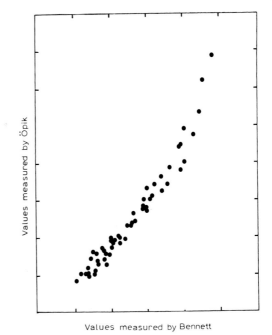

Fig.3. Correlation graph for measurements by ÖPIK (1924) and BENNETT (1938). See caption of Fig.2.

Determination of absolute systematic errors in brightness and surface slope parameters

A method of determining a unique value of the error of any one brightness scale (and hence of all the others) can be developed (WILSON, 1968) by making use of some theoretical photometric function such as that of HAPKE (1966).

Hapke's theoretical equation can be written as:

$$B(\phi, \iota, \varepsilon, f, \gamma, g, k) = A \cdot H(\phi, \iota, \varepsilon, f, \gamma) \cdot S(\phi, k) \cdot R(\phi, g) \tag{2}$$

where A is a normalising constant;

$$H = \frac{q(1-f)\cos \iota}{\cos \iota + \cos \varepsilon} + \frac{f \cdot r \left[\cos(\varepsilon + j\phi)\sin(\gamma + h\phi) - 0.5 \sin^2 \frac{\phi}{2} \ln \left| \frac{\cos(\varepsilon + j\phi) + \sin(\gamma + h\phi)}{\cos(\varepsilon + j\phi) - \sin(\gamma + h\phi)} \right| \right]}{2 \cos \frac{\phi}{2} \cos \varepsilon \sin \gamma} \tag{3}$$

and j, h, q, r take certain quantised values as a function of γ, ι and ε;

$$S = \frac{\sin \phi + (\pi - \phi)\cos \phi}{\pi} + k(1 - \cos \phi)^2 \tag{4}$$

and:

$$R = \begin{bmatrix} 1 - \dfrac{\tan |\phi|}{4g} \{1 - \exp(-g/\tan |\phi|)\} \{3 - \exp(-g/\tan |\phi|)\}, 0 \leqslant \phi \leqslant \dfrac{\pi}{2} \\[2em] 0.5, \dfrac{\pi}{2} \leqslant \phi \leqslant \pi \end{bmatrix} \tag{5}$$

which is a slight generalisation of the original form (WILSON, 1968). Here, R is the retrodirective function, describing the effect of a porous surface with fractional pore space equal to $(g/2)^{\frac{1}{2}}$, and S is a modification of the Schoenberg function which appears adequate to describe the scattering from individual surface fragments. Eq.3 gives the expression H resulting from the integration of the Lommel-Seeliger function, $\cos \iota/(\cos \iota + \cos \varepsilon)$, over a model surface consisting of cylindrical hollows of maximum slope γ occupying a fraction f of a locally flat surface. Analogous functions can be derived using other models of the lunar surface.

The use of such theoretical equations in determining the systematic error relies on the assumption that, for some pairs of values (f, γ), the function H adequately represents the effects of surface slopes. Since the products of R and S has no explicit angular dependence on ι or ε eq.2 can be re-written as:

$$B(\phi, \iota, \varepsilon, f, \gamma, g, k) = H(\phi, \iota, \varepsilon, f, \gamma) \cdot \Phi(\phi, g, k) \quad (6)$$

where B is an error-free brightness measurement.

Some constant error Δ is expected in all observed brightness values B', so that eq.6 becomes:

$$(B' - \Delta) = H(\phi, \iota, \varepsilon, f, \gamma) \cdot \Phi(\phi, g, k) \quad (7)$$

If this equation were to be solved for Φ for all pairs of values (ι, ε) and f and γ, and hence H, were assumed known, it would be found that a constant error Δ would introduce fluctuations in Φ that would vary with both ι and ε but not with ϕ alone. Thus the required value of Δ is that which, on solution of eq.7, leads to values of Φ that vary only with ϕ. In practice, f and γ are not known, so sets of values of Φ have to be generated for permutations of trial sets of values of f, γ and Δ. The set of such values that produces a function Φ, depending as nearly as possible only on ϕ for all ι and ε, represents the best solution to the problem. A suitable test of the independence of Φ on ι and ε is to minimise the function:

equator; the profiles of the models are given in Fig.4. Model 1 is that of Hapke, with cylindrical depressions of maximum slope γ occupying a fraction f of the surface; model 2 has cylindrical ridges of maximum slope Γ occupying a fraction F of the surface; model 3 has rectangular ridges with height to width ratio H occupying a fraction G of the surface; and model 4 consists of infinitely deep, vertically sided clefts occupying a fraction K of the surface.

In Table I the quantities ρ and τ are weighting factors and ρ is a measure of the rate of change of T with Δ near the minimum $T = \tau$. A small value of τ corresponds to a good fit between the data and the theoretical function, and a large value of ρ indicates a sharply defined minimum in T. Values of Δ are, therefore, weighted by ρ/τ; they cluster around a mean of 2.8.

The median slopes of the first two surface models of Table I are, respectively, $5°$ and $12°$ for mare areas and $18°$ and $15°$ for highland areas. These values are

$$T(\Delta, f, \gamma) = \sum_{\phi=-90°}^{+180°} \sqrt{\sum_{\iota=0}^{90°} \mu(\Phi) \left[\left(\frac{\sum_{i=0}^{90°} \Phi(\phi, \iota, \Delta, f, \gamma) \cdot \mu(\Phi)}{\sum_{i=0}^{90°} \mu(\Phi)} \right) - \Phi(\phi, \iota, \Delta, f, \gamma) \right]^2} \quad (8)$$

where ε has been eliminated by use of the relation:

$$\phi = \iota - \varepsilon \quad (9)$$

which holds for points along the lunar brightness equator. Summation is carried out at intervals of $10°$ in ι and $5°$ in ϕ, and the function $\mu(\Phi)$ is such that $\mu = 0$ if no observed value of B' is available but that, otherwise, $\mu = 1$.

The results of this method of finding Δ are given in Table I. The observed values of B' were expressed on the scale of Van Diggelen and four models of the lunar surface were constructed. All models involve ridges or hollows running orthogonal to the brightness

given in Table II together with median slopes determined by radar and infrared methods. Those measurements corresponding to scale lengths of a few centimetres, which are most sensitive to the roughness governing the photometric properties, closely approach the values found in the present analysis. The corrected brightness values, giving the complete photometric functions of the two main types of lunar material, are listed in Table III.

Fig.5 shows graphs of the brightness of typical mare material as a function of ε for three values of ι; this particular set of curves is most useful in the specification of lunar properties when identifying

TABLE I

PARAMETERS OF THE SURFACE MODELS

Model	Number[1]	Mare regions				Continental regions			
		Δ	slope parameters	τ	ρ	Δ	slope parameters	τ	ρ
Cylindrical hollows	1	2.8	$f = 0.22; \gamma = 51.5°$	50.6	6.95	3.2	$f = 1.00; \gamma = 38.0°$	59.3	8.03
Cylindrical ridges	2	1.0	$F = 0.48; \Gamma = 52.0°$	39.9	3.43	3.1	$F = 1.00; \Gamma = 31.0°$	75.5	4.15
Rectangular ridges	3	2.6	$H = 0.0$ for all G	52.0	6.58	2.25	$H = 0.0$ for all G	85.3	3.85
Infinite clefts	4	2.6	$K = 0.0$	52.0	6.58	2.25	$K = 0.0$	85.3	3.85

[1] Refers to Fig.4.

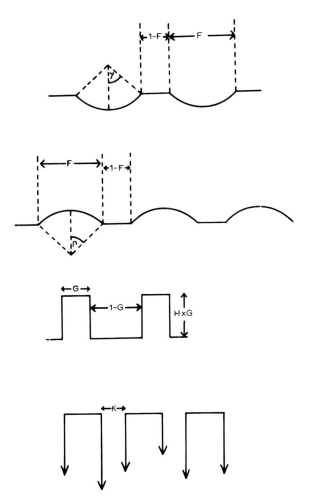

Fig.4. Profiles of the surface models used to deduce slope distributions.

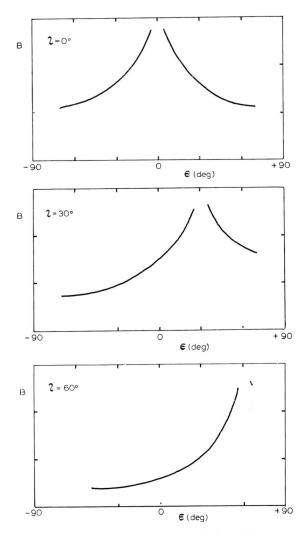

Fig.5. The relative brightness *(B)* of typical marial areas as a function of ε for ι = 0°, 30° and 60°.

possible constituents of the lunar soil (FIELDER et al., 1967). The photometric functions of lunar material are again needed in determining large-scale surface slopes by photoclinometry (WILHELMS, 1963).

The next step is to see how closely the lunar photometric functions conform to Minnaert's rule. Fig.6 shows a plot of a set of points whose co-ordinates are given by:

TABLE II

LUNAR SURFACE SLOPES FROM VARIOUS SOURCES

(Type of) area	Source	Scale length (cm)	Median slope (degrees)
Mare Cognitum	JAEGER and SCHURING (1966)	~300	6
Highlands	SMITH (1967)	100	9
Highlands	EVANS and HAGFORS (1966)	100	10
Average	EVANS and PETTENGILL (1963)	68	5.2
Average	BECKMANN (1965)	3.6	11.2
Average	DANIELS (1963)	3.6	14
Average	EVANS and PETTENGILL (1963)	3.6	8.2
Average	WATSON (1965)	1–10	18
Maria	this chapter	~1	6–12
Highlands	this chapter	~1	15–19

TABLE III

CORRECTED PHOTOMETRIC FUNCTIONS OF MARE AND
HIGHLAND REGIONS

ε (degrees)	ι (degrees)									ε (degrees)	ι (degrees)								
	0	10	20	30	40	50	60	70	80		0	10	20	30	40	50	60	70	80
A. Mare regions										**B. Highland regions**									
−80	—	—	—	—	—	—	—	—	—	−80	22.3	—	—	—	—	—	—	—	—
−75	—	—	—	—	—	—	—	—	—	−75	24.7	22.3	21.7	21.8	—	—	—	—	—
−70	22.8	—	16.8	14.8	—	—	—	—	—	−70	26.8	23.4	21.4	18.7	16.3	—	—	—	—
−65	22.8	20.8	17.4	14.8	12.3	—	—	—	—	−65	29.3	24.2	21.7	19.3	16.8	—	—	—	—
−60	23.4	20.8	17.9	15.3	12.3	10.5	—	—	—	−60	31.8	26.3	22.8	19.8	17.4	14.7	—	—	—
−55	24.3	20.8	18.4	15.3	12.3	10.5	—	—	—	−55	35.2	28.3	24.2	20.3	17.4	14.7	—	—	—
−50	25.3	21.4	18.7	15.8	12.8	10.5	8.3	—	—	−50	37.2	31.2	26.3	21.7	17.8	14.7	12.3	—	—
−45	26.3	22.3	19.5	16.3	12.8	10.5	7.8	—	—	−45	40.7	33.9	27.9	22.3	18.4	15.3	12.3	—	—
−40	27.8	23.8	20.3	17.4	13.3	10.5	7.8	5.3	—	−40	44.3	36.8	31.3	24.2	19.3	15.8	12.3	8.8	—
−35	30.3	25.8	21.4	17.9	13.8	11.5	7.8	4.8	—	−35	47.8	39.3	32.3	26.8	20.3	16.3	12.3	8.8	—
−30	32.8	28.3	22.8	19.5	14.8	11.8	7.8	4.8	3.8	−30	51.2	42.3	35.2	29.3	22.3	17.4	12.3	8.8	—
−25	35.8	30.3	24.7	20.8	15.8	12.3	8.3	4.8	3.3	−25	55.3	46.3	37.8	32.3	24.2	18.4	12.8	8.8	—
−20	38.8	33.3	26.8	22.3	16.8	12.8	8.8	5.3	3.3	−20	59.4	39.9	40.3	34.7	27.4	19.3	13.3	9.3	3.8
−15	42.3	35.8	29.3	23.8	18.4	13.8	9.8	5.8	3.3	−15	63.2	53.3	43.8	37.2	28.8	20.8	14.7	9.3	3.8
−10	47.2	39.9	31.2	25.8	19.8	14.8	10.3	6.3	3.3	−10	69.7	57.2	47.3	39.9	31.2	22.3	15.8	9.8	4.3
− 5	54.8	44.3	33.9	27.8	21.4	16.3	10.5	6.8	3.3	− 5	87.4	62.3	50.4	43.3	34.3	24.2	16.8	10.3	4.3
0	—	47.8	36.3	30.3	22.8	17.4	11.8	7.8	3.3	0	—	65.9	54.3	45.8	36.8	25.8	18.7	10.3	4.8
5	54.8	53.3	39.9	32.3	24.7	18.4	12.8	7.8	3.3	5	87.4	74.3	58.3	49.9	39.9	27.4	20.3	11.4	5.3
10	47.2	—	42.8	34.7	26.8	19.8	14.3	8.8	3.8	10	69.7	—	64.8	53.8	42.3	29.8	21.8	12.3	5.8
15	42.3	48.8	47.2	39.3	29.3	21.8	15.8	9.8	4.3	15	63.2	74.3	74.8	59.3	45.3	32.3	24.2	13.8	6.3
20	38.8	45.3	—	44.8	33.3	23.8	16.8	10.5	4.8	20	59.4	68.8	—	65.3	49.3	34.7	27.4	15.8	6.8
25	35.8	42.3	49.9	51.2	37.8	26.3	18.5	12.3	5.3	25	55.3	64.3	75.2	76.3	55.3	39.3	31.3	17.4	—
30	32.8	38.8	44.8	—	42.3	29.8	20.8	13.8	6.3	30	51.2	59.9	68.8	—	66.4	46.8	33.3	19.8	—
35	30.3	36.3	41.8	52.8	48.3	33.9	23.4	15.8	7.8	35	47.8	56.4	64.3	82.4	74.8	55.3	37.2	—	—
40	27.8	33.3	38.3	46.8	—	37.3	27.5	17.9	9.3	40	44.3	52.3	59.9	—	—	64.8	42.3	—	—
45	26.3	31.2	35.2	42.3	50.5	43.8	31.2	20.3	11.5	45	40.7	48.3	56.4	—	—	—	—	—	—
50	25.3	29.3	33.3	39.3	45.8	—	38.3	23.5	13.3	50	37.2	45.3	52.3	—	—	—	—	—	—
55	24.3	27.8	31.8	36.8	42.3	49.3	49.3	27.4	15.8	55	35.2	41.2	49.3	—	—	—	—	—	—
60	23.4	26.8	29.8	35.2	39.9	46.3	—	33.3	19.8	60	31.8	38.3	46.3	58.3	—	—	—	—	—
65	22.8	26.3	29.3	34.3	38.3	44.8	51.2	43.3	26.3	65	29.3	34.7	42.3	52.3	—	—	—	—	—
70	22.8	25.8	28.3	33.3	36.8	43.8	—	—	35.8	70	26.8	31.3	39.3	46.8	53.8	61.8	—	—	—
75	—	—	—	—	—	—	—	—	43.8	75	24.7	27.4	32.3	42.8	51.2	57.2	—	—	—
80	—	—	—	—	—	—	—	—	—	80	22.3	24.2	—	40.3	49.3	—	—	—	—

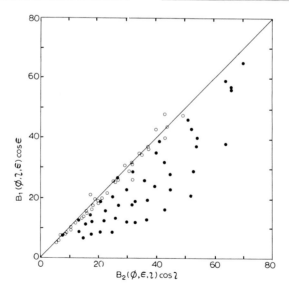

Fig.6. Minnaert's symmetry rule is obeyed by points lying on the straight line. Open circles are obtained from photometric functions derived by the author; filled circles are from ORLOVA's (1956) results.

$$x = B' (\phi, \varepsilon, \iota) \cos \iota \qquad (10)$$

$$y = B' (\phi, \iota, \varepsilon) \cos \varepsilon \qquad (11)$$

According to Minnaert's rule these two expressions should be equal. Hence, the set of points should define a straight line with unit slope and zero intercept. The points generated from the functions presented here scatter around the theoretical line; the points generated from Orlova's tables form the pattern shown in Fig.6. There appears to be systematic error in Orlova's normalisation of Fedoretz's data.

The single particle scattering function and surface porosity

The next phase in the interpretation of the scattering functions already derived is to analyse the function $\Phi\ (\phi, g, k)$. This represents the product of $S(\phi, k)$ and $R(\phi, g)$ given by eq.4 and eq.5, respectively. Sets of values of Φ have been generated in evaluating T from eq.8; the set needed is that corresponding to the best solution for f, γ and Δ.

The functions R and S do not give adequate representation of the lunar properties at $\phi \lesssim 10°$, for there is a brightness surge (GEHRELS et al., 1964) as ϕ approaches zero. HAPKE (1966) proposed that this was due to the porosity of individual fragments. For $\phi \gtrsim 10°$, however, the parameter g, eq.5, will be a measure of first-order porosity and the parameter k, eq.4, will be sensitive to the amount of forward scattering by individual surface particles at $\phi > 90°$.

Since there are no brightness measurements at exactly $\phi = 0°$, the normalising constant A is not known. The following iterative method has been used to determine A, g and k: we have:

$$\Phi (\phi) = A \cdot R (\phi, g) \cdot S (\phi, k) \qquad (12)$$

and:

$$S (\phi, k) = \frac{2}{A} \Phi (\phi), \qquad \phi \geqslant 90° \qquad (13)$$

since:

$$R (\phi \geqslant 90°) = 0.5 \qquad (14)$$

One can adopt the steps:

(1) A first approximation to g is guessed, say g_0.

(2) The resulting value of A, say A_0, is determined from:

$$\Phi (\phi = 10°) = A \cdot R (\phi = 10°, g_0) \cdot S (\phi = 10°, k = 0) \qquad (15)$$

The value of k at $\phi = 10°$ can be neglected since the term $(1 - \cos \phi)^2$ is small.

(3) The first approximation to k, say k_1, is found from eq.13, using A_0.

(4) By analogy with eq.12, a series of values of a function $X(\phi, g)$ are generated at $\phi = 15°, 20°, 25°, \ldots 85°$ with $g = 0.1, 0.2, \ldots 2.0$, such that:

$$X (\phi, g) = \Phi (\phi)/\{S (\phi, k_1) \cdot R (\phi, g)\} \qquad (16)$$

If eq.12 were to hold exactly, for some value of g, $X(\phi, g)$ would be a constant independent of ϕ and equal to A. So:

(5) A test is made of $X(\phi, g)$, at each value of g, by generating the function:

$$U (g) = \sum_{\phi = 15°, 20°}^{80°, 85°} \{\overline{X} (g) - X (\phi, g)\}^2 \qquad (17)$$

where:

$$\overline{X} (g) = \frac{1}{15} \sum_{\phi = 15°, 20°}^{80°, 85°} X (\phi, g) \qquad (18)$$

The function $U(g)$ has a minimum, ideally zero, when $X(\phi, g)$ is constant. Therefore, $U(g)$ is plotted against g and the value, g_1, at which the minimum occurs is noted. Then $\overline{X}(g_1) = A_1$ is the next approximation to A.

(6) The value A_1 is inserted in eq.13 to find a new value of k, say k_2.

(7) Steps (4)–(6) are repeated until no further change takes place in k, g and A. Let the final values be k_n, g_n and A_n, respectively; then:

(8) The final values of S are calculated using $B(\phi, g_n)$ and A_n in eq.12 and eq.13.

With first approximations of $g = 0.7$ and 0.9 for mare and highland areas, respectively, the successive values of g, A and k given in Table IV are found. The

TABLE IV

SUCCESSIVE VALUES OF g, A AND k

Iteration	In maria			In highlands		
	k	g	A	k	g	A
0		0.7	57.36		0.9	83.30
1	0.1171	0.7	57.63	0.1661	0.9	87.74
2	0.1142	0.7	58.02	0.1775	1.0	84.45
3	0.1142			0.1851	1.05	83.85
4				0.1887	1.05	83.70
5				0.1887		

values of g indicate average soil porosities of about 80% for maria and 60% for continental areas.

These values may be compared with: (a) results from the Surveyor craft, which gave estimates of porosities of up to 80% at the surface in Sinus Medii and between 40% and 50% for both types of lunar area at depths of a few centimetres; and (b) results from Apollo 11, which give a lower limit of 46–56% for samples from Mare Tranquillitatis (COSTES et al., 1970).

The single particle scattering functions $S(\phi, k)$ are shown in Fig.7. A greater degree of forward scat-

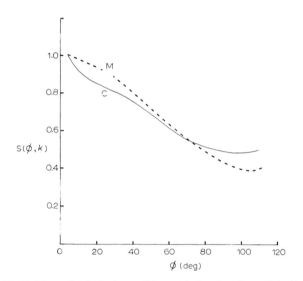

Fig.7. Normalised single particle scattering functions $S(\phi, k)$ for average marial *(M)* and highland *(C)* terrain.

tering is shown by particles in continental regions.

Since, in the present analysis, the photometric properties of many lunar areas of similar but not necessarily identical morphologic types have been averaged, the values of the surface parameters deduced here form useful working values but are of limited accuracy. The analyses could be repeated to greater effect if accurate brightness measurements were available for a single lunar area at all angles ι, ε. If the deduced photometric parameters of some small areas were correlated with a detailed examination of the same areas through the use of a suitably instrumented soft landing probe, a technique for rapidly surveying the surface properties of any atmosphereless planet would be available using orbiting probes.

We now return to the problem of the shape of the photometric function at small angles of vision. GEHRELS et al. (1964) made measurements at angles of vision as small as 0.8° and, by extrapolation, obtained a brightness surge of 85%, from $\phi = 5°$ to $\phi = 0°$, which did not depend appreciably on the type of lunar terrain. POHN et al. (1969) have measured an Apollo photo-

graph to obtain data actually at $\phi = 0°$, and the brightness surge is found to be only 44% from $\phi = 5°$ to $\phi = 0°$.

Several explanations for this "opposition effect" have been proposed. HAPKE (1966) suggested that individual fragments of the regolith themselves consisted of aggregates of smaller particles so that their effective porosity was possibly as much as 99.5%. GEHRELS et al. (1964) prefer a suspension of interplanetary particles supported by electrostatic charges as suggested by GOLD (1955), though this model is not consistent with the shape of the photometric function at large angles of vision. Certainly the scale size of fragments responsible for the effect must be very small, since the opposition brightening is seen in photographs taken from as little as 1 m above the surface.

Neither of the above suggestions is necessarily in conflict with the findings from the Apollo 11 mission. A large fraction of the regolith, when sieved, consists of fragments less than 50 μ in diameter, and there is evidence for some electrostatic effects (COSTES et al., 1970). A suspension of micron-sized particles[1] would not persist on samples removed from the lunar surface and transported to Earth. However, a brightness surge at small angles of vision may be seen on some terrestrial rocks (OETKING, 1965). A complete explanation of the opposition effect has yet to be given.

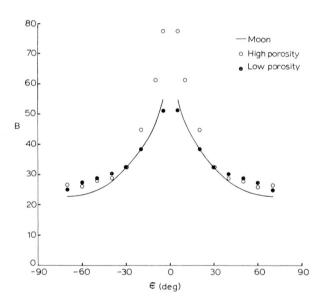

Fig.8. Comparison of photometric curves: brightness *(B)* as a function of ε for $\iota = 0°$ of: *(a)* Moon observed from Earth (curve for maria); *(b)* Apollo 11 fines sample in a state of low porosity (filled cirles); and *(c)* Apollo 11 fines samples in a state of high porosity (open circles).

---

[1] A dust detector has been used on the Apollo 12 mission to measure such particles in situ.

It is important to know to what extent changes have occurred in the physical structure of regolith samples during the return to Earth and, to this end, the photometric curves of Apollo 11 fines have been determined in the laboratory. Fig.8 shows the brightness curves for $\iota = 0$ of such a sample in two stages of roughness and compaction, compared with the corresponding curve, taken from Fig.5, for lunar maria.

As expected, the sample will match the lunar average for some intermediate state of packing.

The photometric techniques outlined in this chapter, especially when combined with polarimetric and thermal measurements, can provide a significant amount of information about the physical properties of a planetary surface.

References

BARABASHEV, N. P., 1922 *Astron. Nachr.*, 201: 289.

BECKMANN, P., 1965. *J. Geophys. Res.*, 70: 2345.

BENNETT, A. L., 1938. *Astrophys. J.*, 88: 1.

COSTES, N. C., CARRIER, W. D., MITCHELL, J. K. and SCOTT, R. F., 1970. *Science*, 167: 739.

DANIELS, F. B., 1963, *J. Geophys. Res.*, 68: 449.

EVANS, J. V. and HAGFORS, T., 1966. *J. Geophys. Res.*, 71: 4871.

EVANS, J. V. and PETTENGILL, G. H., 1963. *J. Geophys. Res.*, 68: 423.

FEDORETZ, V. A., 1952. *Uch. Zap. Kharkov Univ.*, 42:49.

FIELDER, G., GUEST, J. E., WILSON, L. and ROGERS, P. S., 1967. *Planetary Space Sci.*, 15: 1653.

GOLD, T., 1955. *Monthly Notices Roy. Astron. Soc.*, 115: 585.

GEHRELS, T., COFFEEN, T. and OWINGS, D., 1964. *Astron. J.*, 69: 826.

HAMEEN-ANTTILA, K. A., 1967. *Ann. Acad. Sci. Fennicae, Ser. A VI*, 252.

HAPKE, B. W., 1966. *Astron. J.*, 71: 333.

JAEGER, R. M. and SCHURING, D. J., 1966. *J. Geophys. Res.*, 71: 2023.

MARKOV, A. V., 1924. *Astron. Nachr.*, 65: 221.

MARKOV, A. V., 1950. *Bull. Abastumani*, 11: 107.

MINNAERT, M. J. G., 1941. *Astrophys. J.*, 93: 403.

OETKING, P., 1965. *Trans. Am. Geophys. Union*, 46: 131.

ÖPIK, E., 1924. *Publ. Astron. Obs. Tartu.*, 26: 1.

ORLOVA, N. S., 1956. *Astron. Zh.*, 33: 93.

POHN, H. A., RADIN, H. W. and WILDEY, R. L., 1969. *Astrophys. J.*, 157: L193.

ROSENBERG, H., 1921. *Astron. Nachr.*, 214: 137.

SAARI, J. M. and SHORTHILL, R. W., 1967. *Boeing Sci. Res. Lab., NASA Contr. Rept.*, 855.

SHARANOV, V. V., 1939. *Uch. Zap. Leningr. Gos. Univ.*, 31: 28.

SINTON, W. M., 1967. *Icarus*, 6: 222.

SMITH, B. G., 1967. *J. Geophys. Res.*, 72: 4059.

VAN DIGGELEN, J., 1958. *Rech. Astron. Obs. Utrecht*, 14(2).

WATSON, K., 1965. *Astrogeol. Ann. Study Program Rept.*, A, 55.

WILHELMS, D. E., 1963. *Astrogeol. Ann. Study Program Rept.*, D, 1962–1963, 1.

WILSON, L., 1968. *Interpretation of Lunar Photometry*. Thesis, Univ. London, London, pp.29, 41, 71, 180.

WIRTZ, C., 1915. *Astron. Nachr.*, 201: 289.

9

Polarimetric studies

E. L. G. BOWELL

Introduction

A study of the degree of polarisation of the top surface of the lunar regolith affords a powerful method of determining some of the physical properties associated with the fine particulate material and large rocks exposed to sunlight.

Before lunar materials were returned to Earth polarimetric studies had been used to ascertain that almost all the exposed lunar surface was composed of rocks that resembled fine-grained terrestrial basalts. In addition, experiments on terrestrial rocks had pointed to the nature of some of the environmental processes that had altered the lunar surface. It had been shown, for example, that the bombardment of terrestrial rock samples by protons and α-particles brought their polarimetric properties closer to those of the Moon.

The results of polarimetric measurements on the fines and consolidated rocks returned to Earth by the Apollo 11 mission proved to be startlingly similar to the results of Earth-based telescopic measurements on the Moon. Since this returned material was grossly disturbed and churned following its collection, these results are most important in indicating that the polarimetric properties of the lunar surface itself may be essentially the same as those of the subsurface material.

For comprehensive surveys of lunar polarimetry the reader may refer to LYOT (1929), DOLLFUS (1961), or PELLICORI (1969). Some of the techniques and results of the lunar polarisation programme carried out at the Paris–Meudon Observatory by DOLLFUS and BOWELL (1969) and DOLLFUS et al. (1970), are described below. This was essentially a twin study: measurements of selected areas on the Moon's surface were made at the telescope, and terrestrial simulations (pseudo-lunar materials) were used in the laboratory to interpret the lunar results.

The lunar polarisation curve

Light reflected obliquely from the surface of the Moon is partially plane-polarised. The azimuth of the largest component lies either in, or perpendicular to, the plane of vision containing the Sun, the point observed on the Moon and the observer. If I_1, I_2 are the respective components of the light intensity perpendicular to, and in the plane of vision, then the degree of polarisation is defined by:

$$P = \frac{I_1 - I_2}{I_1 + I_2} \tag{1}$$

A polarisation curve describes, for a given lunar region, the degree of polarisation as a function of the angle of vision, ϕ (Chapter 8). Fig.1 shows two typical lunar polarisation curves drawn from measures made in orange light of wavelength $0.60\,\mu$ with a Lyot visual fringe polarimeter. Curve a refers to a marial region of normal albedo $A = 0.063$ in Oceanus Procellarum; curve b to a highland region centred on the crater Krüger ($A = 0.117$). It can be seen that, at full Moon ($\phi \approx 0°$), the degree of polarisation, P, is zero. For angles of vision $0° < \phi \lesssim 23°$, P is negative, and $P_{\min}$ is -1.2% at $\phi(P_{\min}) \approx 11°$. For larger angles of vision P increases and becomes positive at the inversion angle $\phi(0) \approx 23°$. Subsequently P increases to a maximum at $\phi(P_m) \approx 100°$. For $\phi(P_m) < \phi < 180°$ P decreases, tending to zero at $\phi = 180°$.

A lunar polarisation curve may therefore be considered in two parts:

(1) The negative branch, $P < 0$, for which $0 < \phi \lesssim 23°$: at a given wavelength the form of the negative branch is similar for all lunar terrains (marial or highland). Likewise P_m is constant for all regions, changing slightly with wavelength from -1.0% at $0.325\,\mu$ to -1.2% at $1.05\,\mu$. The angle of inversion increases systematically with wavelength, from $20.7°$ at $0.325\,\mu$

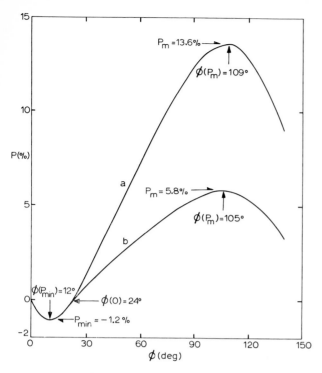

Fig.1. Lunar polarisation curves in orange light ($\lambda = 0.60\ \mu$). The degree of polarisation, P, is plotted against the angle of vision, ϕ, for two regions: *(a)* in Oceanus Procellarum (near Grimaldi); and *(b)* Krüger.

to 26.2° at 1.05 μ. The form of the negative branch characterises the physical microstructure of the regolith; it may be taken that the invariance of the negative branch for the entire lunar earthside disk is indicative of extreme textural uniformity of the optically reflective layer.

(2) The positive branch, $P > 0$, for which $23° \lesssim \phi < 180°$. The following remarks may be made about the characteristics of the positive branch in orange light alone: *(a)* regions of low albedo (maria or mare-like crater floors) display a high P_m of up to 14.0% at $\phi(P_m) \approx 105°$; *(b)* regions of high albedo (highland regions; some stratigraphically young craters) display a low P_m, sometimes as small as 3% (Aristarchus), for which $\phi(P_m) \approx 98°$; and *(c)* there is no dependence of P on the inclination of the macrosurface or position on the lunar disk.

Furthermore, observations made at several wavelengths lead to the following comments:

(a) P_m is greater in the blue than in the red. The relation between P_m and wavelength appears to be similar for all regions.

(b) P_m is likewise greater for regions of low albedo at any particular wavelength in the range of wavelengths used (0.325–1.05 μ). The highest value noted for the maximum polarisation is 35.0% for a region

in Oceanus Procellarum, observed at a wavelength of 0.325 μ. The lowest value measured is 3.5% for a continental region northeast of Mare Crisium, observed at 1.05 μ.

(c) Although the angle of vision $\phi(P_m)$ corresponding to the maximum polarisation varies with P_m it is independent of wavelength.

Polarisation measurements on lunar regions

Three polarimeters were employed in the observing programme at the Meudon and Pic du Midi Observatories between 1965 and 1967. One of these is the classic instrument designed by Bernard Lyot (LYOT, 1929; DOLLFUS, 1961), a visual fringe polarimeter used by eye and capable of isolating lunar areas as small as 2 arcsec diameter. The other two polarimeters are photoelectric: one for the infrared region of the spectrum up to 1.1 μ, described by MARIN (1965); the other for the ultraviolet and visible spectrum from 0.32 to 0.6 μ. All three instruments have a measuring capability better than 0.1% in P.

The observations may be split into two parts: *(1)* polarisation curves as complete as possible were determined for fourteen large (65 arcsec diameter) lunar regions in eight colours covering the whole wavelength range of the polarimeters; and *(2)* the degrees of polarisation of 142 small (< 5 arcsec diameter) regions were measured in orange light at lunar visual angles near 100°—corresponding to the lunar phase angle of maximum polarisation.

Fig.2, 3 and 4 illustrate some of the data collected from the study of large regions. In Fig.2, which refers to a region of low albedo in Oceanus Procellarum, it may be seen that the maximum polarisation, which occurs at a visual angle of about 110°, is very great in the ultraviolet, diminishing as the wavelength increases to become quite small in the infrared. Fig.3 shows similar curves for Mare Crisium. The albedo of this region is higher than that of Oceanus Procellarum, and all the values of maximum polarisation are smaller. Fig.4 relates to a highland region of fairly high albedo near the crater Krüger. The maxima of these curves are again lower than in the preceding case. All the negative branches display similar forms whilst the inversion angle varies systematically with wavelength as shown in Fig.5.

Following the work of WRIGHT (1935), MARKOV (1958), and WILHELMS and TRASK (1965), the study of 142 small regions was undertaken in order to correlate the maximum degree of polarisation with

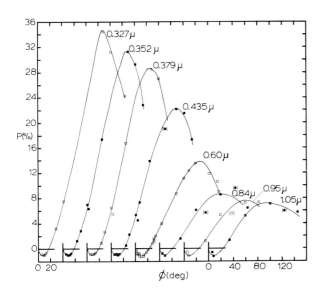

Fig.2. Polarisation curves in eight colours for a region in Oceanus Procellarum (near Grimaldi) of 65 arcsec diameter. The origins of curves have been staggered across the diagram for clarity; the interval in angle of vision is 20°. Poor measures are bracketed.

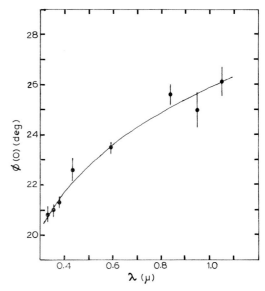

Fig.5. Variation of the inversion angle, $\phi(0)$, of lunar polarisation curves with wavelength, λ, taken from measurements on fourteen regions of 65 arcsec diameter.

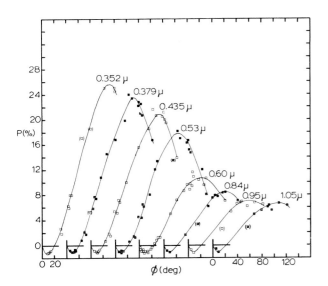

Fig.3. Polarisation curves (caption as in Fig.2) for Mare Crisium.

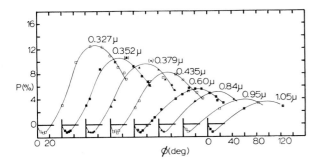

Fig.4. Polarisation curves (caption as in Fig.2) centred on the crater Krüger.

albedo and stratigraphy. A wide variety of lunar terrains was selected to cover the entire range of both major stratigraphic units and albedo in the hope that anomalies in maximum polarisation might uncover some geologic or mineralogic peculiarities.

Many authors, for example HAPKE (1964) and WEHNER et al. (1963), have constructed diagrams of A versus P_m which show a domain within which all lunar measurements lie. A logarithmic plot, constructed from measurements made in orange light on 67 small regions, shows that a linear relation exists between $\log A$ and $\log P_m$ (Fig.6):

$$\log A = -0.724 \log P_m - 1.808 \qquad (2)$$

If the magnitudes of the estimated errors shown in Fig.6 are correct it seems likely that the linear relation between $\log A$ and $\log P_m$ applies to all lunar terrains.

It is desirable to extend eq.2 to other wavelengths in order to more fully characterise the relation between A and P_m for the lunar surface. The observations made on fourteen large regions are suitable for the evaluation of P_m at eight wavelengths; but data concerning the spectral variation of the normal albedoes of these regions are not available. It is, however, possible to determine the spectral variation of A and P_m for the whole earthside lunar disk. From the observations of fourteen large lunar regions P_m is found to vary with wavelength as follows:

$$\log P_m = -1.137 \log \lambda + k \qquad (3)$$

where k depends on the terrain type and λ is the wavelength expressed in microns. Taking the mean lunar

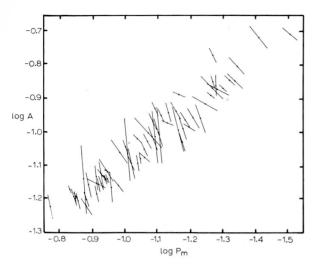

Fig.6. Relation between log A and log P_m. The vectors give the estimated error limits. Marial regions appear at the bottom left and highlands regions at the top right.

normal albedo at $\lambda = 0.60\ \mu$ to be $\bar{A} = 0.101$, eq.2 may be used to derive the maximum polarisation for the earthside disk: $\bar{P}_m = 0.0766$.

Then, according to eq. 3, the spectral variation of the maximum degree of polarisation for the earthside disk is given by:

$$\log \bar{P}_m = -1.137 \log \lambda - 1.396 \qquad (4)$$

Measurements of normal albedoes have been made by GEHRELS et al., (1964, table XIII) for thirteen small lunar regions in four colours spanning the wavelength range 0.36–0.94 μ. Analysis of these results gives the albedo of the earthside disk at various wavelengths:

$$\log \bar{A}_\lambda = 0.823 \log \lambda - 0.797 \qquad (5)$$

Combining eq.4 and eq.5 leads to a relation which describes the spectral variation of the maximum degree of polarisation and albedo:

$$\log \bar{P}_m = \log \bar{A}_\lambda - 1.96 \log \lambda - 0.60 \qquad (6)$$

which may be written approximately ($\pm 5\%$, λ being again in microns) as:

$$\bar{P}_m = \frac{\bar{A}_\lambda}{3.5\ \lambda^2} \qquad (7)$$

Table I lists values of the maximum polarisation, $\bar{P}_m$, and normal albedo $\bar{A}_\lambda$ of the whole earthside disk of the Moon at various wavelengths, λ.

In summary, it will be noted that the emphasis of the observational programme has been on the positive branch of the lunar polarisation curve and, in particular, on the evaluation of the maximum degree of polarisation and its spectral variation for a number

TABLE I

MAXIMUM POLARISATION AND ALBEDO OF THE EARTHSIDE DISK OF THE MOON AS A FUNCTION OF WAVELENGTH

$\bar{P}_m$	A_λ	λ (μ)
0.169	0.060	0.327
0.155	0.067	0.353
0.141	0.074	0.377
0.121	0.087	0.434
0.077	0.101	0.600
0.061	0.137	0.830
0.051	0.153	0.950
0.044	0.160	1.050

of lunar regions. This has been done for two reasons:

(1) P_m may be accurately (and fairly easily) evaluated using a small number of repeated readings on a single occasion, provided that the lunar phase angle is not greatly different from the phase angle at which the polarisation attains its maximum.

(2) P_m varies greatly for different lunar terrains and rapidly with wavelength. A study of the forms of the negative branch reveals no such variations in the polarimetric signature of any lunar feature. The reason is that the negative branch arises from absorption, refraction and multiple scattering within the optically active layer of the lunar regolith which, on a kilometre scale, seems to have been altered widely by environmental factors (see Chapter 10) and to have attained a uniform texture.

Polarimetric properties of terrestrial rocks and meteorites

Following the series of lunar measurements made at Meudon between 1965 and 1967 an investigation was instigated into the principal polarimetric properties of some 60 terrestrial rock and meteorite samples. The purpose of this study was to evaluate in a general way the relation between P_m and A for specimens ground to powders with average grain sizes ranging from 160 μ down to 26 μ in diameter. In this manner it was possible to propose certain rock types as constituents of the lunar regolith, and to investigate the texture and median grain size of the surface layer and the effects of some of the denudation processes that operate on the Moon. The rock samples were divided into three groups: (1) eleven basic and intermediate samples collected from sites in Hawaii by G. Fielder (see FIELDER et al., 1967); (2) a group of twenty samples, comprising a wide variety of igneous rocks, collected by J. E. Guest in Chile and A. Dollfus in Arizona; and some Hawaiian volcanic ashes supplied by D. Coffeen; and (3) a variety of meteorites supplied mostly by the Muséum National d'Histoire Naturelle, Paris. Some micrometeoritic dust collected by S. Grjebine in Italy was also analysed.

DOLLFUS (1961) showed that the degree of polarisation of rock samples measured in the laboratory should be related to the angles of incidence and emergence as well as to the over-all inclination of the sample's surface to the plane of vision. However, for finely ground powders of most terrestrial rocks it is sufficient to measure the degree of polarisation as a function of the angle of vision only. In the Meudon study polarimetric measures were made at angles near those of specular reflection.

Fig.7 shows polarisation curves for some of the Hawaiian rocks from group (1). These samples were ground and sifted to pass a 200-μ mesh. Some of the curves (notably curves 3 and 6) resemble the lunar polarisation curves of Fig.1 quite closely; they have well developed negative and positive branches similar in form to that of a typical lunar curve. However, the curves representing most terrestrial rocks do not display lunar-like characteristics. Fig.8 illustrates the results of measurements on some acidic samples from

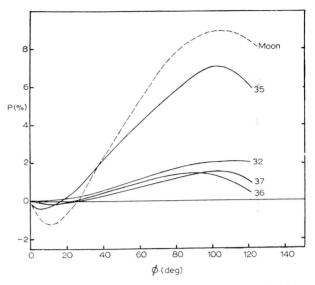

Fig.8. Polarisation curves in orange light ($\lambda = 0.60\ \mu$) for four terrestrial samples (35 = diorite; 32 = granodiorite; 37 = granite; 36 = syenite) which do not match the lunar polarisation signature. A typical lunar curve is shown for comparison.

group (2). Here the negative branches are almost non-existent and the maximum degrees of polarisation for three of the four samples are depressed far below any lunar values.

A plot of normal albedo versus maximum degree

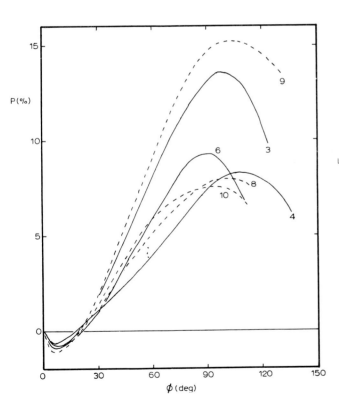

Fig.7. Polarisation curves for six numbered Hawaiian rock samples sieved to pass a 200-μ mesh. Measured in orange light of wavelength 0.60 μ

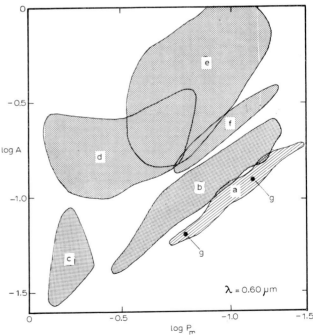

Fig.9. Schematic classification of some terrestrial substances and meteorites according to albedo A and maximum degree of polarisation, P_m: a = lunar measures; b = fine-grained basic and intermediate rocks; c = pulverised vitric basalt; d = crushed coarse-grained rocks; e = sands, clay, chalk; f = meteorites; g = bronzite–chondrites (two points only).

of polarisation has been constructed to assist in the classification of terrestrial rocks. Fig.9 shows schematically how various classes of pulverised terrestrial material occupy different positions in the diagram. The lunar measurements, taken from Fig.6, occupy a narrow domain (a) in the lower right of Fig.9. It is quite clear that certain substances such as sands, clays and chalks (e), crushed coarse-grained rocks (d), and pulverised vitric basalt (c) cannot comprise the lunar regolith. A remarkable place is occupied in the diagram by the two bronzite–chondrites (g) from Ochansk and Pultusk. The physical properties of these meteorites will have to be compared in more detail with those of lunar materials. The other meteorites studied (f) are of interest since they fall in a narrow region similar in extent to the one occupied by the lunar measurements, though their albedoes are somewhat higher than those of the lunar areas.

Only the fine-grained basic and intermediate rocks, (b) justify a detailed comparison with the lunar measurements. Although the agreement is not perfect, errors in the photometric scales (lunar and laboratory) or systematic modification of the optical properties of rocks at the lunar surface might cause a slight shift of the domain b from the lunar domain.

The effects of grain size and possible lunar surface denudational processes were investigated. Reducing the median grain size of a rock sample almost invariably reduces its maximum degree of polarisation and increases its albedo without significantly affecting the shape of the negative branch of its polarisation curve. Fig.10 illustrates these effects for a very fine-grained basalt from group (1). Here, the median grain sizes of the sieved powders are 160, 83, 62 and 26 μ. These effects may also be shown on a plot of log A versus log P_m as in Fig.11; all the Hawaiian rocks from group (1) have been ground and sieved into the four grain sizes above (only grain sizes of 160 and 26 μ are illustrated). Points corresponding to the larger grain size are to the lower left of the diagram. The shift to the upper right for smaller grain sizes is, in most cases, accompanied by an approach to the domain containing the lunar results, especially for the intrinsically opaque rocks (sample numbers 1, 3 and 9, for example). If points for each of the four grain sizes are plotted separately to form domains on a log A versus log P_m diagram it becomes clear that the smaller grain sizes are those that approach the lunar results most closely. By inspecting the positions of these domains a median grain size of 23 $\pm$ 3 μ may be attributed to the lunar regolith. PELLICORI (1969) had concluded that the lunar surface consisted of particles 20–35 μ in diameter,

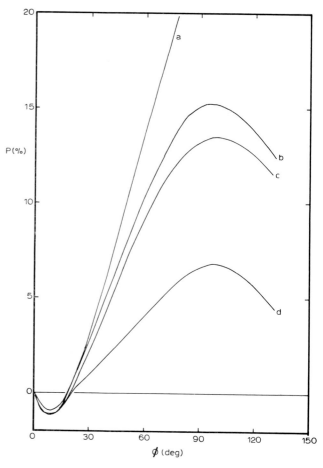

Fig.10. Polarisation curves of sample GF/H10 in orange light. Median grain sizes are: (a) 160 μ; (b) 83 μ; (c) 62 μ; and (d) 26 μ.

and that these particles were somewhat translucent.

Reducing the grain size of certain materials can lead to polarisation curves closely approaching those of the Moon. However, highly pulverised materials

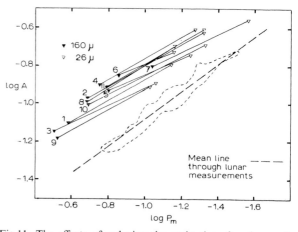

Fig.11. The effects of reducing the grain size of rock samples from group (1). The broken curve encloses results from lunar measurements; some of the sample numbers correspond to those in Fig.7.

generally have albedoes which exceed those common to the Moon. Two other processes may exert a major influence on the sculpturing of the regolith and, hence, on the lunar polarimetric properties:

(*1*) Darkening by solar wind particles: HAPKE (1965, 1966) irradiated silicate powders with 2 keV protons and noted a systematic darkening for most. DOLLFUS and GEAKE (1965) irradiated a single chondritic meteorite and noted a change in polarimetric properties as well as a reduction of albedo (fig.2 in Chapter 10).

(*2*) Upwelling in vacuo: FIELDER et al. (1967) exposed molten rocks at about 1,200 °C to vacuum conditions and found that violent release of gases from the rocks was accompanied by darkening and autobrecciation—spontaneous fracturing and ejection of small particles during cooling.

Fig.12 shows schematically, in a log A versus log P_m plot, how the alteration of a fresh rock exposed

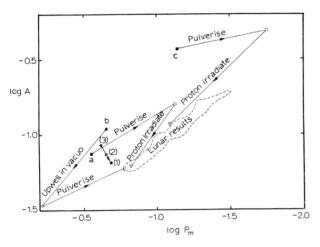

Fig.12. Possible schemes for the alteration of rocks on the lunar surface: *a, b* = basalts; *c* = dunite. *(3) → (2) → (1)* brecciation of crystalline lunar material (see lunar surface model described in text).

on the lunar surface may proceed under the influence of the processes described above. For the basaltic sample, *a*, lunar properties may be simulated by first reducing the grain size and afterwards irradiating the sample with protons. For the basalt, *b*, the lunar domain can be reached by upwelling first and reducing the grain size afterwards. In the same way the dunite sample, *c*, can be pulverised and subsequently irradiated with protons in order to enter the lunar domain.

Fig.12 may give the impression that the observed lunar polarisation values can be reproduced with a large range of different samples; but, in fact, the process has severe limitations. Many of the lunar polarimetric

properties are not catered for in these examples: the simulated material has to exhibit a lunar-like negative branch and spectropolarimetric properties similar to those of the Moon. The conclusion is that it is possible to achieve a perfect match between certain, limited terrestrial rock samples and the Moon by applying to these rock samples the actual environmental processes that occur on the Moon.

Investigation of returned lunar samples

Polarisation measurements on materials returned by Apollo 11 were made by A. Dollfus, J. E. Geake, C. Titulaer and R. J. Fryer at the Meudon Observatory, Paris, and reported by GEAKE et al. (1970). Three samples were studied: (*1*) a small quantity of fines; (*2*) a chip of crystalline material from the surface of a very dark grey rock; and (*3*) a chip of crystalline material from the surface of a rock containing much light-toned material.

The polarisation curve for the fines was found to be identical with the curve from Earth-based observations of Mare Tranquillitatis. Not only are the negative branches of the same form, but the spectral variation of P_m is also identical within the limits of experimental error. Fig.13 illustrates the results graphically; the solid curves represent Earth-based measurements of a large area in Mare Tranquillitatis. Dollfus and his co-workers noticed that, when the surface of the fines sample was smoothed and the material thereby slightly compacted, the negative branch became slightly shallower.

The chip of dark, crystalline material displayed a slightly less developed negative branch than the fines, extending only to an angle of vision of about 19°. The maximum degree of polarisation on the other hand, attained the very high value of about 35% at an angle of vision of about 130°—greater than that found from Earth-based observations of the Moon. The light, crystalline material quite uncharacteristically revealed no negative branch; whilst the positive branch attained a peak similar to that of the dark, crystalline material.

GEAKE et al. (1970) subjected the fines sample to irradiation by 60 keV protons for 2 h (a dose of about 1 $\mu a/cm^2$). There was no change in either its albedo or its polarisation signature. Continued irradiation for 7 h caused no detectable changes.

On heating the fine material for a short time, to 500 °C or more, a colour change from dark grey to a lighter, more reddish colour was noticed; subsequent proton irradiation caused perceptible darkening but no change in the polarisation curve.

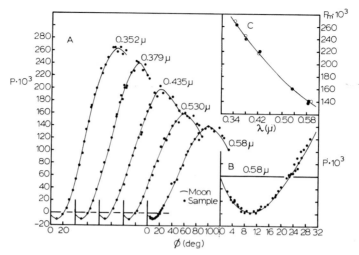

Fig.13. Polarisation curves for Mare Tranquillitatis (solid curve; open circles) and returned lunar fines (filled circles). A. Complete curves in five colours. B. The negative branch on a larger scale. C. The change of P_m with wavelength. (After GEAKE et al., 1970.)

The principal significance of the agreement between Earth-based lunar observation and measurements on the returned fines is that the optical properties of the exposed lunar surface seem to be the same as those of the subsurface regolith—at least down to a depth of some centimetres. From the lack of change of the optical properties of the fine material after its irradiation with protons it might be inferred that the fines have been damaged, to saturation, throughout the depth sampled. The polarisation curves for the two chips of crystalline rock, dissimilar to those of the fines and to the Earth-based lunar curves, suggest that nowhere on the lunar surface can such clean rocks comprise a large fraction of the exposed material.

Consider a model of the lunar surface comprising a mosaic of fines and exposed rock fragments having polarimetric properties similar to those measured by Dollfus. Let the respective fractions exposed be Q_1 and Q_2, their albedoes A_1 and A_2, and the corresponding maximum degrees of polarisation P_{m_1} and P_{m_2}, where the subscript 1 refers to fines, 2 to the solid rock. Then the albedo of the mosaic is given by:

$$\bar{A} = Q_1 A_1 + Q_2 A_2 \qquad (8)$$

and its maximum degree of polarisation by:

$$\bar{P}_m = \frac{Q_1 A_1 P_{m_1} + Q_2 A_2 P_{m_2}}{Q_1 A_1 + Q_2 A_2} \qquad (9)$$

Suppose that the fraction Q_2 of exposed rock is 1/10 and that its maximum degree of polarisation P_{m_2} is 0.35. Consider rocks of three albedoes: (1) $A_2 = 0.2$

(dark crystalline chip); (2) $A_2 = 0.3$; (3) $A_2 = 0.4$ (light crystalline chip). The calculated values of $\bar{A}$ and $\bar{P}_m$ have been plotted in Fig.12. If denudation were to follow the sequence (3)→(2)→(1), then it might be possible for rocks such as the light crystalline chip to be incorporated into the lunar regolith by progressive darkening.

This scheme is a gross simplification of the structure and evolutionary pattern of the regolith since the proportion of exposed, consolidated rock cannot remain constant over long periods of time; yet it does serve to show that rocks such as the light crystalline chip cannot comprise significant fractions of any part of the lunar surface.

Now it is inferred from a study of thermal anomalies (Chapter 11) in such regions as the crater Tycho (Chapter 5) that outcrops of large rock blocks occupy up to 10% of the exposed surface. Evidently, if the thermal and polarimetric results for Tycho are to be compatible we must expect either that the surfaces of these rocks are darkened by solar wind irradiation or that they are coated with finer, darker material.

Finally, it is interesting to note that the median particle size (weighted by mass) of the returned lunar fines is about 40 μ (HAPKE et al., 1970) and that over 35% of the mass is contained in particles smaller than 20 μ. According to DUKE et al. (1970) the lunar fines are notably deficient in material finer than 15 μ, compared with selected terrestrial samples. They estimate a median grain size (weighted by mass) of 62 μ and a modal grain size (weighted by number) of 20 μ.

References

DOLLFUS, A., 1961. In: G. P. KUIPER and B. M. MIDDLEHURST (Editors), *Planets and Satellites*. University of Chicago Press, Chicago, Ill., 3: chapter 9.

DOLLFUS, A. and BOWELL, E. L. G., 1969. *Final Report, 1.* U.S.A.F. Contract AF 61 (052)-914, European Office of Aerospace Research.

DOLLFUS, A. and GEAKE, J. E., 1965. *Compt. Rend.*, 260: 4921.

DOLLFUS, A., BOWELL, E. L. G. and TITULAER, C., 1970. *Final Report, 2.* U.S.A.F. Contract AF 61 (052)-914, European Office of Aerospace Research.

DUKE, M. B., CHING CHANG WOO, BIRD, M. L., SELLERS, G. A. and FINKELMAN, R. B., 1970. *Science*, 167: 648.

FIELDER, G., GUEST, J. E., WILSON, L. and ROGERS, P., 1967. *Planetary Space Sci.*, 15: 1653.

GEAKE, J. E., DOLLFUS, A., GARLICK, G. F. J., LAMB, W., WALKER, G., STEIGMENN, G. A. and TITULAER, C., 1970. *Science*, 167: 717.

GEHRELS, T., COFFEEN, D. and OWINGS, D., 1964. *Astron. J.*, 69: 826.

HAPKE, B., 1964. In: J. W. SALISBURY and P. E. GLASER (Editors), *The Lunar Surface Layer*. Academic Press, New York, N.Y.

HAPKE, B., 1965. *Ann. New York Acad. Sci.*, 123: 711.

HAPKE, B., 1966. In: W. N. HESS, D. H. MENZEL and J. A. O'KEEFE (Editors), *The Nature of the Lunar Surface*. Johns Hopkins, Baltimore, Md., p.143.

HAPKE, B., COHEN, A. J., CASSIDY, W. A. and WELLS, E. N., 1970. *Science*, 167: 745.

LYOT, B., 1929. *Ann. Obs. Meudon*, 8(1). (See also: *NASA Tech. Transl.*, F-187, 1964.)

MARIN, M., 1965. *Rev. Opt.*, 44: 115.

MARKOV, A. V., 1958. *Izv. Gl. Astron. Obs.*, 158: 138.

PELLICORI, S. F., 1969. *Opt. Sci. Center, Univ. Arizona (Tucson), Tech. Rept.*, 42.

WEHNER, G. K., ROSENBERG, D. L. and KENKNIGHT, C. E., 1963. *Planetary Space Sci.*, 11: 1957.

WILHELMS, D. E. and TRASK, N. J., 1965. *Astrogeol. Studies*, A: 63.

WRIGHT, F. E., 1935. *Ann. Rept. Smithsonian Inst.*, 168. See also: G. P. KUIPER and B. M. MIDDLEHURST (Editors), *Planets and Satellites*. Univ. Chicago Press, Chicago, Ill., 4: chapter 1.

10

Chemical problems of the surface

ALISON P. BROWN and L. WILSON

Bulk chemistry and rare elements

Prior to the Surveyor landings in 1967 and 1968 the chemical nature of the lunar surface was unknown. It was commonly argued that lunar rocks would not be compositionally dissimilar to terrestrial basalts, though a few scientists had expected either highly acidic rocks or meteoric material to predominate. Chondrites are the most common type of meteorite, probably accounting for over 90% of all meteorites. A chondritic Moon would have lent support to the theory which assumes that all the planets were formed by an aggregation process.

Surveyor 5, using an alpha-scattering device, produced the first elemental analyses (Chapter 5, table II) of the lunar surface in the southern part of Mare Tranquillitatis. The alpha particles were produced from about 100 millicuries of Curium-242 in the sensor head which rested on the lunar surface. TURKEVICH et al. (1968) concluded that lunar mare rock was not dissimilar to basaltic achondrites and terrestrial basalts. The results (Chapter 5, table II) obtained from Surveyor 6, which landed in Sinus Medii, paralleled those drawn from Surveyor 5.

Surveyor 7 landed on the rim unit of Tycho at 40.95°S 11.41°W. The analyses (Chapter 5, table II) were similar to those drawn from previous Surveyors except that there was a lower concentration of the Fe-group elements (mass numbers 47–65). Thus it is possible that the lower albedo in the mare regions is partly due to opaque minerals, such as ilmenite ($FeTiO_3$), which are included in the Fe-group.

The lunar rocks collected from Mare Tranquillitatis by the crew of Apollo 11 (July 1969) consisted of five main types: 57.4% by volume of the samples was soil breccia, 37.4% basalts, 3.6% anorthosites, 5.1% glasses and the remaining 1.5% was meteoric material or shocked debris. In general terms lunar rocks are of a uniform composition and the differences between them arise chiefly in the concentrations of the minor and trace elements potassium, rubidium, caesium, uranium, thorium and barium. All lunar rocks have an unusually high concentration of titanium, scandium, zirconium, hafnium, yttrium and tri-valent rare earths. They are low in sodium and, surprisingly, in europium, relative to the other rare earths (Fig.1). The high titanium and low europium contents suggest that the lunar rocks are the results of a long, fractional crystallisation process. Both the ^{87}Rb–^{97}Sr, and ^{40}K–^{40}Ar age determinations indicate that the crystalline rocks returned from Mare Tranquillitatis solidified about $3.7 \cdot 10^9$ years ago.

The soil and breccia compositions are similar to those of the rocks, but the soil is enriched in nickel and other volatiles (cadmium, zinc, silver, gold, copper and thallium) which occur in carbonaceous chondrites. In fact, the enhancement of these elements in the soil is consistent with the addition of such meteoric material. Some authors (for example, ALBEE et al., 1970) suggest that the correlation of barium with alkali metal concentration, the anticorrelation of strontium with alkalies, and the almost constant proportion of calcium, are probably the result of magmatic processes. However, the enhancement, relative to chondritic material, of lithium and gadolinium in the lunar rocks may reflect a difference in original composition. Water and ferric oxides are, as generally expected, absent in lunar rocks (PECK and SMITH, 1970) but halogen concentrations are similar to those in meteorites. MOORE et al. (1970) report weighted mean abundances of carbon and nitrogen in the lunar fines of 140–255 and 100–150 p.p.m., respectively. The finest sieved fraction contained most carbon and nitrogen. As contaminants, both elements could be indigenous, meteoric, solar or terrestrial in origin. It is difficult to determine the amount of carbon deposited from the solar wind since

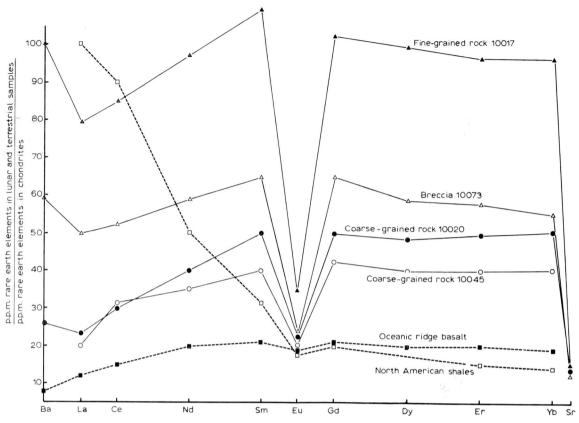

Fig.1. Europium deficiency of Apollo 11 lunar samples compared with two terrestrial rock types. (After GAST and HUBBARD, 1970; HASKIN et al., 1970.)

the "sticking" factor for carbon is not known; Moore et al. estimate that about 4% of lunar carbon is from this source. Meteoric material probably accounts for 0.5–2.0% of carbon by weight. In each case the amount of nitrogen from these sources is about 20% by weight that of the carbon. Indigenous nitrogen may be more important than carbon if these estimates are accurate.

KAPLAN and SMITH (1970) determined carbon and sulphur concentrations in six samples of fines, breccia and rock and obtained results between 20–200 and 650–2,300 p.p.m. respectively. The fines and breccia contained a higher proportion of carbon than the whole lunar rock but a lower proportion of sulphur. In the fine-grained rocks, the concentration of carbon varied between 20 and 60 p.p.m.; that of sulphur lay between 2,200 and 2,280 p.p.m. In the bulk fines carbon concentrations were about 150 p.p.m.; sulphur concentration was about 640 p.p.m. The fine-grained rocks had the lowest carbon content and contained the lighter carbon isotope. This may indicate that those samples containing only traces of lunar carbon are most influenced by terrestrial contamination. Similarly the results for sulphur suggest that the fines are enriched with ^{34}S relative to the rocks. The fines are also

enriched with ^{13}C. Elemental sulphur was not found; all the sulphur was apparently present as the sulphide. Processes operating on the surface either preferentially enrich the fines with heavier sulphur and carbon isotopes (relative to the lunar rocks, terrestrial volcanic rocks, and meteoric material) or remove the lighter ^{12}C and ^{32}S. Kaplan and Smith prefer the latter explanation, suggesting stripping of S and C by solar wind protons which favours the conversion of ^{12}C to methane and ^{32}S to hydrogen sulphide. Table I and II give specimen analyses of Apollo 11 rocks, for minor and trace elements. Table I shows results obtained by X-ray fluorescence analyses, and Table II extends these analyses to include other elements using isotope dilution and atomic absorption techniques.

No important organic molecules have been found in lunar rocks or fines but ABELL et al. (1970) reported that some proportion of the gaseous hydrocarbons released from the fines on treatment with aqueous HCl, H_3PO_4 or HF were indigenous. Up to 6 μg/g of CH_4 and other gaseous paraffins (up to C_4H_{10}) have been released from the fines. Abell et al. further suggest that the hydrocarbons present might arise from hydrolysis of carbides (which probably originated in meteoric

TABLE I

X-RAY FLUORESCENCE SPECTROGRAPHIC ANALYSES
(p.p.m.) OF LUNAR SAMPLES

	Igneous rocks[1]		Breccia (10018*)	Soil (10084*)
	group I (10072*)	group II (10045*)		
Ba	300	10	175	134
Rb	5.6	0.6	3.6	2.96
Sr	168.4	137.7	158.5	164.8
Pb	3.4	3.3	4.2	2.9
Th	3.5	0.4	2.2	2.5
Zr	497	194	328	318
Nb	25	14	19	18
Y	162	73	106	99
La	43	16	24	21
Ce	94	32	67	58
Pr	16	6	11	10
Nd	49	17	29	33
V	22	98	51	36
Cr	2,280	2,400	1,950	1,850
Co	30	23	35	34
Ni	20	20	200	230
Cu	22	20	32	33
Zn	34	14	54	37
Ga	4	3	4	4

[1] There are two groups of igneous rocks distinguished by chemical analysis: group I is higher than group II in Rb, K, Ba, Th, Ti, Ni, P, S, Zr, Y and rare earths, but lower in Al and Ca.
* Houston sample designation. (Adapted from COMPSTON et al., 1970.)

TABLE II

TRACE ELEMENTS (p.p.m.) IN LUNAR SAMPLES

	Analytical error $(\pm\%)^1$	Fine-grained rocks (10017*)	Coarse-grained rocks (10020*)	Breccia (10073*)	Soil (10084*)
Na	10(AA)	3,800	2,800	3,500	3,200
K	3(ID)	2,610	486	1,200	1,200
Rb	3(ID)	5.63	0.63	2.9	2.79
Cs	10(ID)	0.155	—	0.098	0.104
Sr	2.4(ID)	174.8	149.8	167.5	171.4
Ba	5(ID)	308	77.1	175	176
La	3(ID)	26.2	7.95	—	16.3
Ce	3(ID)	77.2	25.8	46.5	47.6
Nd	3(ID)	59.7	24.1	35.4	36.8
Sm	3(ID)	21.0	9.49	12.4	13.1
Eu	3(ID)	2.15	1.61	1.70	1.77
Gd	5(ID)	27.0	12.8	16.9	17.2
Dy	5(ID)	34.1	16.2	19.2	19.7
Er	5(ID)	19.2	9.99	11.5	12.1
Yb	5(ID)	18.8	9.77	10.7	11.5
Lu	7(ID)	2.66	1.43	1.56	1.58
Ni	15(XRF)	17	10	161	198
Zn	15(AA)	30	10	24	24

[1] (ID) = analysis by stable isotope dilution; (AA) = atomic absorption analysis; (XRF) = X-ray fluorescence analysis of powder.
* Houston designations of samples. The analytical error represents 90% confidence limits. (Adapted from GAST and HUBBARD, 1970.)

material) by OH-groups formed from solar wind proton bombardment, from direct contamination of C and H from the solar wind, or by thermal or radiation alteration of indigenous organic compounds.

The crew of Apollo 12 landed on Oceanus Procellarum in November 1969. The rocks at this site are similar in appearance to those at the Apollo 11 site; however, a preliminary report prepared by the LUNAR SAMPLE PRELIMINARY EXAMINATION TEAM (1970) points out several interesting differences. The Apollo 12 rocks appeared to be about 10^8 years younger than the Apollo 11 samples; about half of the material collected was microbreccia whereas this accounted for only 2% of the Apollo 11 material; and the regolith at the Apollo 12 site appeared to be only half as thick as at the Apollo 11 site. The fines from Oceanus Procellarum are lighter in tone than those from Mare Tranquillitatis and contain different proportions of phases. Pyroxene, plagioclase and olivine are found as before and glass accounts for about 20% of the sample. Chemically, although both Apollo sites show the expected high concentrations of refractory elements and low concentrations of volatiles, there are differences between the two sites. Apollo 12 samples (fines and rocks) have a lower titanium content (1.2–5.1% TiO_2 compared with 7–12% in Apollo 11 samples) and also a lower concentration of potassium, rubidium, zirconium, yttrium, lithium and barium. The crystalline rocks from Apollo 12, however, have a greater concentration of iron, magnesium, nickel, cobalt and vanadium. The Lunar Sample Preliminary Examination Team note that the significant difference between the rocks from the two sites is among the elements which favour ferromagnesian minerals: the range of abundance is much less in the Apollo 12 samples. Total carbon analyses show that, as in the Apollo 11 samples, the highest carbon abundances are found in the fines (up to 200 p.p.m.). A total carbon abundance of about 40 p.p.m. appears to be characteristic of lunar rocks. Apparently there are no detectable organic molecules in Apollo 12 samples—concentrations must be less than 200 parts per 10^9. Preliminary conclusions on the Apollo 12 site suggest that it is less mature, geomorphologically, than the Apollo 11 site, and that the rocks there more nearly resemble terrestrial tholeiitic and alkaline basalts. Their lower content of solar wind gases confirms that the Apollo 12 rocks have been exposed to space for a shorter time than the rocks in Mare Tranquillitatis.

The effects of radiation on surface chemistry

The surface of the Moon, unprotected by an atmosphere, is continuously bombarded with high-energy particles and radiation. In considering chemical changes, it is justifiable to investigate only the effects of proton bombardment. High-energy electromagnetic radiation may cause re-arrangement of chemical bonds and probably some lattice damage; but it does not affect the bulk chemical composition. Neutrons, which are also incident on the lunar surface, can cause nuclear reactions but their incidence is too low to produce significant chemical changes. High-energy alpha particles also contribute to nuclear reactions occurring on the lunar surface; on capturing electrons, low-energy alphas (which constitute about 15% of the solar wind) become relatively inactive helium atoms.

The protons incident on the lunar surface have energies ranging from 1 keV to 1 geV or greater. The protons of lowest energy are part of the solar wind from a quiescent sun. The flux of protons of energy 1–100 keV has been estimated (ZELLER et al., 1970) to be 10^{16} cm^{-2} per year.

HAPKE (1964) and WEHNER et al. (1965) suggested that low-energy solar wind protons caused darkening of the lunar surface layer. The greatest depth of penetration of these protons in solid material is one or two microns but Hapke and Wehner et al. conclude that exposure of rock samples for (lunar) times of 10^5 years leads to a significant modification of the surface properties.

The apparatus used by Hapke to test this suggestion was based on a design by Kaufman. A hydrogen plasma was maintained in one chamber by electron bombardment of the gas. Ions were extracted through a grid and accelerated to 2 keV into the main chamber where they bombarded the target material. The ion current density was 1 ma/cm^2 and the beam contained H_2^+ and H_3^+ ions as well as protons. The darkened samples have polarimetric and photometric properties (Fig.2) that more nearly approach those of the lunar surface than those of the original, unbombarded material.

Hapke's results were supported by Wehner et al., who used an ion source consisting of an RF-excited plasma, in which the sample was immersed. Again, the incident flux contained a high proportion of H_2^+ and H_3^+ ions; the ion current density was about 2 ma/cm^2. Both of these systems utilised oil diffusion pumps backed by oil-sealed mechanical fore-pumps to produce a vacuum inside the chamber. Operating pressures were between 10^{-3} and 10^{-6} torr. Both

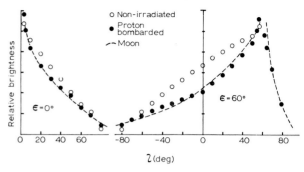

Fig.2. Photometric properties of irradiated and non-irradiated basalt powders compared with the Moon. (After HAPKE, 1966.)

Hapke and Wehner et al. report that alpha particles produce effects similar to those induced by protons, though with greater efficiency.

In contradiction to these results, NASH (1967) suggested that darkening occurred in laboratory experiments as a result of contamination and excessive sample temperature. His results, obtained using a RF-excited hydrogen plasma to produce mostly H_1^+ ions and a cold cathode getter-ion pump, indicated that, although darkening of the sample occurred, the darkening was due to contamination rather than surface damage. He showed that the darkening is both dose-dependent and dose-rate dependent and also that it increases as the temperature of the sample's surface is raised. He also noted a difference in the amount of darkening occurring on basalt and quartz powders, a result reported by both Wehner et al. and Hapke. Hapke used this result to argue that darkening was not due to contamination since it varied with the chemical composition of the sample, but Nash suggested that the result could be explained by catalytic action. Silica-magnesia and silica-alumina are good promoters of hydrocarbon decomposition while high concentrations of sodium and potassium act as inhibitors. Basaltic rocks may, therefore, be more efficient catalysts than acidic rocks which contain a higher percentage of alkali metals.

Nash also noted that the darkening spread over the edge of the sample cups and suggested that this was further evidence for contamination. The source of the contamination could be organic volatiles from the oil-sealed mechanical fore-pump or from the Viton "O" rings in the system. These sources of contamination were present in the vacuum chambers used by Hapke and Wehner et al. and were also present in Nash's system where the "rough" pumping was done by an ordinary oil-filled rotary pump. Further, Nash pointed out that proton bombardment in solar wind simulation experiments caused a rapid rise in the surface temperature of the sample when the beam power exceeded 0.1 W/cm². The use of high-intensity ion beams to simulate long-term lunar effects must be subjected to careful scrutiny, since laboratory bombardment produces an artificially high temperature.

The problem of discriminating between true surface damage and accidental organic contamination can be solved only by using equipment that is completely free from organic material. To this end, one of us (WILSON, 1968) built an RF-excited gas source to provide ions for an electrostatic linear accelerator (Fig.3).

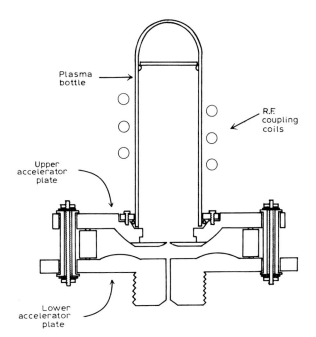

Fig.3. Scheme of 2 keV accelerator used to simulate the solar wind.

A plasma is excited in the appropriate gas (hydrogen and helium to deliver protons and alpha particles, respectively) at a pressure of about 0.01 torr by coupling the radiation from an RF oscillator with the gas. The oscillator is of the Colpitts type, providing 200 W at about 70 MHz. A borosilicate glass bottle contains the plasma and is directly mounted above a hole in the upper of the two circular plates that form the electrostatic accelerator. The required potential is maintained across the plates, which are shaped (with azimuthal symmetry about a vertical axis) in such a way that the field focusses ions into a narrow beam passing through a canal in the lower plate.

The ion beam then coasts into the high-vacuum sample chamber, a pressure gradient being maintained across the canal by differential pumping. Contamination of the target area is eliminated by the use of

modern vacuum pumps: the roughing pumps are of the molecular sieve sorbtion type, cooled with liquid nitrogen, and the main vacuum pump is an orbitron, described by MALIAKAL et al. (1964). The other common source of contamination—vacuum "O" rings—is eliminated by the use of all-metal seals: indium seals at all glass to metal contacts and gold seals elsewhere. With the input power mentioned above, a beam power density of 0.4 W/cm² for 2 keV protons can be attained.

In a preliminary experiment, a soda glass sample was bombarded with 1.8 keV protons for 12 h with a beam current of 20 μA, corresponding to an exposure time of about 3,000 years on the Moon; there was no indication of darkening. The beam power density in this experiment was 0.06 W/cm², below Nash's threshold for excessive heating.

Earlier investigation of the darkening effect produced by 60 keV protons on a powdered sample of soda-glass showed that, using a system with an oil-filled roughing pump, three hours bombardment with a beam current of 3 μa/cm² provided visible darkening. Mass spectrometric analysis (BROWN, 1970) of this sample, together with an unbombarded but otherwise similar sample, demonstrated the existence of a peak at 105 ± 2 mass units in the bombarded sample only. This peak first appeared at an analysis temperature of 190°C and was not seen at temperatures above 600°C. The peak was almost certainly due to hydrocarbon contamination and could have been a fluorinated fragment derived from the Viton "O" rings which contained a fluorinated polymer. GREER and HAPKE (1967) reported some rather inconclusive microprobe analyses of powders darkened by proton bombardment. They detected carbon and iron impurities on the darkened samples but stated that it was impossible to subscribe to the idea that these two elements were solely responsible for the darkening.

Analyses of the Apollo rocks have not, so far, shed much light on the problem of solar wind darkening. The bulk of the evidence suggests that the darkening found in the laboratory is due to both hydrocarbon and metal contamination. Contamination of this type in vacuum systems in which charged particles bombard a target has been known for many years. WATSON (1947) described contamination of targets in the beam of an electron microscope. Work by ENNOS (1953) shows that the rate of contamination increases with current density up to 100 ma/cm²; but also that high sample temperatures eliminate contamination.

In view of the suggestion by NASH (1968) that the original samples (which darkened during bombardment) used by Hapke and Wehner et al. probably reached high temperatures, it is interesting to observe that Ennos was dealing primarily with hydrocarbon contamination, whereas that due to metals and to fluorocarbons would be more difficult to remove by heating.

Deposition of carbon from the solar wind has also been proposed as a possible darkening process occurring on the lunar surface. To what extent carbon from the solar wind would "stick" to the lunar surface is not known, so it is not possible to calculate the solar contribution to the carbon in the lunar surface material. MOORE et al. (1970) have suggested that about 4% of the carbon found on the Moon is solar wind deposited; the maximum amount of carbon found on Apollo 11 fines is only 200 p.p.m. so that this process is probably not of major importance in terms of the lower albedo of marial rock compared with that of the lunar highlands.

Protons with energies higher than 100 keV have a lower average flux than protons from the solar wind. Those with energies between 100 keV and 10 MeV arrive at the lunar surface in numbers between 10^{12} and 10^{13} cm^{-2} per year. These protons may produce chemical change and damage in the top few millimetres of the lunar surface.

When a high-energy proton penetrates the surface, it immediately looses energy. Elastic and inelastic collisions occur with the lattice atoms, causing them to be ionised. Thermal excitation of the lattice will also decelerate the incoming particle. High, very localised, temperatures ("temperature spikes") will be produced for very short periods of time, "freezing in" the resulting lattice disorder. At the end of its track, the proton will probably capture an electron and remain as an interstitial hydrogen atom. It may also form a chemical bond.

Silicates bombarded with 1 MeV protons show a characteristic absorption in the infrared, at 2.8 μ, due to OH radicals (ZELLER et al., 1970)—a reaction confirmed by the use of deuterons for the bombardment and the detection of the OD absorption at 3.7 μ. In oxygen-rich rocks, the most probable reaction is that of the incoming proton with an oxygen atom. ZELLER et al. have also experimented with the bombardment of graphite and diamond and, although their work is not conclusive, they have suggested that H–C bonds are produced in bombarded graphite, and that small quantities of hydrocarbons are formed having molecular weights of up to 78.

Using electron spin resonance analysis on rock samples bombarded with hydrogen ions Zeller et al. have shown that protons and deuterons are retained in a diamond lattice, although not as interstitial atoms. (The characteristic H and D electron spin resonance spectra are absent.) They suggest that the atoms are retained as hydride ions which may then react with atoms in the surrounding lattice.

The estimated influx of protons with energies ranging from 10 MeV to 1 GeV is 10^{10}–10^9 cm^{-2} per year and that of particles with energies above 1 GeV (protons and heavy nuclei) is about 10^8 cm^{-2} per year. Although these particles will penetrate 1 m or more into the surface and cause nuclear reactions, their flux is so small that the resulting chemical changes would be essentially undetectable, even with present-day techniques.

Volcanism and surface chemistry

Chemical processes on the lunar surface are less complex than those occurring under terrestrial conditions. The main chemical changes on the surface of the Moon are due to two main factors: particle bombardment and sublimate deposition. At the present time the Apollo analyses have been insufficient to resolve the question of the precise role of solar wind bombardment on the lunar rocks. Once a better knowledge of the over-all composition of the lunar surface has been gained it should be possible to describe in detail the reactions which have occurred.

In certain areas of the lunar surface, where volcanism (Chapter 2–5) has played a major role, proton bombardment may cause other chemical bonds to be formed. With the lack of an oxidising atmosphere, volcanic sublimates produced on the Moon will be dissimilar to terrestrial sublimates.

The sublimates on Mount Etna, a fairly typical basaltic stratovolcano, consist primarily of alkali metal oxides, chlorides and sulphates. These are formed from reactions between the more volatile components of the lavas and the gases from the magma. Although, in the initial stages of the reaction, the gases, (hydrogen sulphide, carbon monoxide, and hydrogen, for example) are of a reducing character, they soon react with the Earth's atmosphere to produce oxides (sulphur dioxide, sulphur trioxide, water and carbon dioxide). Whereas, on the Earth, the sublimates and fumarolic products of a volcano are the result of reactions between magmatic gases and our own atmosphere, on the Moon the oxidation of magmatic gases and other products will not occur and the resulting sublimates will probably be in a reduced form.

In the case of terrestrial lava flows the alkali

metals, sodium and potassium, are volatile and escape from the molten rock. Under vacuum conditions, as on the Moon, a higher percentage of these metals will be lost, in the form of the pure metal, from vents and flows. NAUGHTON et al. (1965) suggested that the vapour of the pure metal might act as a degenerative agent on lunar surface material, breaking down rocks into fine powders. Fields of fumaroles are a characteristic of most terrestrial volcanoes. Oxidised gases percolate upwards through layered materials and initiate various chemical reactions. Cool fumaroles emit mainly carbon dioxide and steam, hotter ones may produce quantities of sulphurous gases. The carbon dioxide and water are almost certainly atmospheric in origin.

Volcanism on the Moon must differ in several important ways from that on the Earth. Given the same initial physical conditions, de-gassing processes on the Moon must be more violent than those on the Earth. Excessive temperatures were not necessarily prevalent on the surfaces of lunar lava flows and lakes. The surface temperatures of many terrestrial flows and lava lakes are several hundred degrees higher than the temperatures below the surface simply because of the exothermic oxidation reactions occurring on the surface. With temperatures characteristic of terrestrial flows at depth in the melt, temperatures of up to 1,200°C—but not as high as the 1,400° or 1,500°C measured on some terrestrial flows—might be expected at the surface of a fresh lunar flow. The more quiescent activity associated with terrestrial volcanoes (fumarolic activity) may not exist as such on the Moon; there, the more violent de-gassing, together with the lack of an oxidising atmosphere, may lead to a situation where alkali metals and reducing gases percolate rapidly through the ash and lava layers. Deposits of sulphur and alkali sulphides may be more important than they are on Earth. "Wood's Spot", near Aristarchus, has been suggested as an area with some deposited sulphur. Since one does not expect strong oxidising gases to be formed at the lunar surface, reactions with the lunar rocks are likely to be much less important than in the case of their terrestrial counterparts. The passage of

the alkali metals through the lava layers may cause degeneration of the rock, as Naughton et al. suggested. Bombardment of volcanic areas with protons will, in general, cause little change in the over-all chemistry, which will be that of a strongly reducing environment. S-H bonds may be produced and SH radicals may become of some importance in an area where sulphur is a more plentiful—or, at least, more easily accessible—atom than the oxygen locked in the rocks.

As far as the minor trace elements are concerned, lunar volcanic processes will probably not provide the numerous "carrying" gases and liquids which redistribute these elements in terrestrial fumarolic areas. Many metals are leached from the rocks by the acid gases and solutions which percolate upwards and are then deposited, as the temperature drops, in zones around the fumarole. A common example of this, in many terrestrial volcanoes, is the transport of silicon by volcanic gases. The silicates of the rock are attacked by the fluorine of hydrogen fluoride (HF), at temperatures in excess of 400°C, to produce gaseous silicon tetrafluoride (SiF_4). This compound is then hydrolysed at lower temperatures (100°C or less) by the following reaction:

$$SiF_4(gas) + nH_2O \rightleftharpoons SiO_2(n-2)H_2O + 2H_2SiF_6$$

producing the deposits of silica which are commonly found around high temperature fumaroles.

Even if the original reaction to silicon tetrafluoride occurred on the Moon this gas would be lost: the net result would be a depletion of silicon around volcanic vents. This depletion would extend also to those metals (for example, iron and aluminium) which form fairly volatile chlorides. Fluorine and chlorine and their products—generally the first and most volatile products of a volcanic eruption—will, most probably, be lost quickly into space.

Even terrestrial volcanic processes still hold many secrets and it will surely be many years before we know, from first-hand knowledge, as much about lunar volcanism as we now know about terrestrial magmatic processes.

References

ABELL, P. I., EGLINGTON, G., MAXWELL, J. R., PILLINGER, C. T. and HAYES, J. M., 1970. *Nature*, 226: 251
ALBEE, A. L., BURNETT, D. S., CHODOS, A. A., EUGSTER, O. J., HUNEKE, J. C., PAPANASTASSIOU, D. A., PODOSEK, F. A., PRICE RUSS II, G., SANZ, H. G., TERA, F. and WASSERBURG, G. J., 1970. *Science*, 167: 463.
BROWN, A. P., 1971. *Physical and Chemical Studies of the Lunar Surface*. Thesis, University of London, London, Chapter 5.

COMPSTON, W., ARRIENS, P. A., VERNON, M. J. and CHAPPELL, B. W., 1970. *Science*, 167: 476.
ENNOS, A. E., 1953. *Brit. J. Appl. Phys.*, 4: 101.
GAST, P. W. and HUBBARD, N. J., 1970. *Science*, 167: 485.
GREER, R. T. and HAPKE, B. W., 1967. *J. Geophys. Res.*, 72: 3131.
HAPKE, B. W., 1964. *Centre Radiophys. Space Res., Cornell Univ., Rept.*, 169.
HAPKE, B. W., 1966. In: W. N. HESS, D. H. MENZEL and J. A.

O'KEEFE (Editors), *The Nature of the Lunar Surface.* Johns Hopkins Press, Baltimore, Md., p.143.

HASKIN, L. A., HELMKE, P. A. and ALLEN, R. O., 1970. *Science,* 167: 487.

HONDA, F. and MIZUTANI, Y., 1968. *Geochem. J.,* 2: 1.

KAPLAN, I. R. and SMITH, J. W., 1970. *Science,* 167: 541.

LUNAR SAMPLE PRELIMINARY EXAMINATION TEAM, 1970. *Science,* 167: 1325.

MALIAKAL, J. C., LIMON, P. J., ARDEN, E. D. and HERB, R. G., 1964. *J. Vac. Sci. Tech.,* 1(2).

MOORE, C. B., LEWIS, C. F., GIBSON, E. K. and NICHIPORUK, W., 1970. *Science,* 167: 495.

NASH, D. B., 1967. *J. Geophys. Res.,* 72: 3089.

NAUGHTON, J. J., BARNES, I. L. and HAMMOND, D. A., 1965. *Science,* 149: 630.

PECK, L. C. and SMITH, V. C., 1970. *Science,* 167: 532.

TURKEVICH, A. L., FRANZGROTE, E. J. and PATTERSON, J. H., 1968. *Science,* 162: 117.

WATSON, J. H. L., 1947. *J. Appl. Phys.,* 18: 153.

WEHNER, G. K., KENKNIGHT, C. E. and ROSENBERG, D. L., 1965. *Ninth Quarterly Status Report.* NASA Contract 751, Appl. Sci. Div., Litton Systems, Inc.

WILSON, L., 1968. *Interpretation of Lunar Photometry.* Thesis, University of London, London, p.143.

ZELLER, E. J., DRESCHOFF, G. and KEVAN, L., 1970. *Mod. Geol.,* 1: 141.

11

Thermal studies

J. A. BASTIN AND E. L. G. BOWELL

Review of the thermal properties of the lunar surface[1]

Introduction

The plane homogeneous model of the lunar surface with temperature-independent thermal constants gives a good approximation to the observed results and the theory is of great heuristic value in deducing the nature and cause of the various anomalies which have been found from thermal observations of the lunar surface layer. Any thermal observation that is not explicable on the basis of the simple model may be related to a thermal anomaly. Adaptation of the plane homogeneous model to fit the observations either locally or on a global scale can provide insight into the nature of the anomaly and this can lead to a better understanding of the physical properties of the lunar crust as well as of the Moon as a whole.

The plane homogeneous model

Suppose that the density, ρ, specific heat, c, and thermal conductivity, k, are independent of the temperature specified by the symbol T_z on the absolute scale, z being the depth below the lunar surface. Expressions may be derived for T_z as a function of z during both eclipse and lunation conditions. Without introducing appreciable error, except perhaps near to the poles, one may assume that the Moon rotates at constant distance for the Sun, at constant synodic angular velocity, ω, and about an axis which is normal to the plane defined by the lunar orbit (see, for example, ALLEN, 1962). The thermal diffusion equation can be written in terms of only one spatial coordinate, z, and the time, t:

$$\frac{\partial T_z}{\partial t} = \frac{k}{\rho c} \left(\frac{d^2 T_z}{dz^2} \right), \quad z > 0 \tag{1}$$

Solutions of this equation will be subject to the boundary condition at the surface that the heat absorbed from the Sun will be equal, by energy conservation, to the sum of the heat radiated from the surface and the heat conducted into the Moon's interior:

$$(1 - \alpha_1) I = (1 - \alpha_2) \sigma T^4 - k \left(\frac{\partial T}{\partial z} \right)_{z=0} \tag{2}$$

where I is the insolation or rate at which solar energy reaches unit area of the lunar surface, σ is Stefan's constant and α_1 and α_2 are the respective mean albedoes for incoming solar radiation and for the longer wavelength thermal radiation emitted from the Moon. The factor $(1 - \alpha_1)$ is introduced since some of the radiation from the Sun is reflected or scattered at the lunar surface. The factor $(1 - \alpha_2)$ represents the amount of radiation radiated by the lunar surface as a fraction of that radiated by a black body at the same temperature. Neither $(1 - \alpha_1)$ nor $(1 - \alpha_2)$ differ significantly from 0.90 (HARRIS, 1961; MURRAY, 1965) and their omission from eq.2 does not greatly affect the solution. The insolation I is related to the solar constant f and to the solar zenith angle ξ by the expressions:

$$
\begin{aligned}
I &= f \cos \xi & 0 < \xi < \pi/2 \text{ (lunar day)} \\
I &= 0 & \pi > \xi > \pi/2 \text{ (lunar night)}
\end{aligned}
\tag{3}
$$

The solar zenith angle can be expressed (see, for example, SMART, 1960) as a function of latitude, ψ, longitude, ϕ and the time t:

$$\cos \xi = \cos \psi \cos (\omega t - \phi) \tag{4}$$

During an eclipse the insolation is given by:

$$I' = \beta I \tag{5}$$

---

[1] This section is written by J. A. Bastin.

where β is a function of the calculated selenographic latitude and longitude and of the time and, ideally, should take account of solar limb darkening.

Eq.1–5 may be solved to give the temperature as a function of depth throughout a lunation and during eclipse. Although, in the case of a lunation, series solutions have been derived (JAEGER, 1953a, b) and homological reasoning is also a valuable tool (WESSE-LINK, 1948), eq.1–5 can be solved relatively simply using high speed computers. The variations of temperature throughout a lunation and during an eclipse, at the lunar surface and at various depths below the surface, are shown in Fig.1–4. The quantity $(k\rho c)^{-\frac{1}{2}}$—the thermal inertia—appears as a variable parameter in the calculations.

hence higher values of ω' in the harmonic analysis of the boundary input conditions.

Infrared observations in the 8–14 μ range may be used to deduce the thermodynamic surface temperature T_0. If F_λ is the flux observed, then, at any given wavelength:

$$F_\lambda = (1 - \alpha_\lambda) B_\lambda (T_0) \qquad (6)$$

where $B_\lambda (T_0)$ is the Planck (black body) function at temperature T_0 and α_λ the albedo at that wavelength. Using this equation (or an integrated extension of it over a known wavelength range) and known values of α_λ for terrestrial or lunar rocks (HARRIS, 1961; MURRAY, 1965; ADAMS and JONES, 1970) the temperature T_0 can be found. Temperatures determined in this way in-

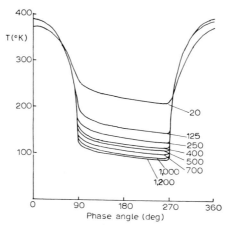

Fig.1. Surface temperature, T, at a point on the lunar thermal equator, computed as a function of phase angle for various values of the inertia parameter γ. The surface of the Moon as a whole corresponds approximately to the value $\gamma = 1,000$ although there are areas— e.g., Tycho—for which the effective value of γ is about 125 whilst spectral thermal evidence for these areas points to a composite surface with whole rocks having γ of about 20. (After KROTIKOV and SHCHUKO, 1963.)

It is clear from Fig.2 that the amplitude of the temperature wave decreases with depth. This result follows simply from eq.1, which cannot be satisfied by an undamped, progressive wave. Similarly, the coefficient of attenuation with depth increases with frequency of oscillation of the surface temperature. This can be seen by applying a simply harmonic temperature boundary condition, of frequency ω', to eq.1. The analysis indicates attenuation of the temperature wave with depth characterised by the coefficient $(\omega'\rho c/2k)^{\frac{1}{2}}$. This result has two important effects: first, during a lunation, the temperature variation becomes smoother as the depth increases since the high harmonics (high ω') are more rapidly damped than the fundamental; second, the thermal disturbance caused by an eclipse does not propagate to such great depths as the effects of the lunation cycle since the eclipse is associated with a much shorter period and

dicate a thermal inertia of 10^3 cm$^2 \cdot$ sec$^{\frac{1}{2}} \cdot$ (K$^\circ$)$^{+1} \cdot$ cal.$^{-1}$ for both lunation and eclipse measurements and the variation of the temperature with time, in both cases, agrees to a first approximation with that predicted from the smooth homogeneous model. With measured values of the specific heat, 0.06–0.02 cal $\cdot$ g$^{-1} \cdot$ (K$^\circ$)$^{-1}$ (ROBIE et al., 1970) and density, 1.58–1.70 g/cm^3 (FRYXELL et al., 1970) of lunar rocks it is clear that the mean conductivity of the regolith close to the lunar surface is about $5 \cdot 10^{-6}$ cal./cm $\cdot$ sec $\cdot$ K$^\circ$. This is in fair agreement with the direct measurements of conductivity (Table I) so far made on lunar material.

Before discussing refinements of this simple picture it will be as well to consider the far infrared and microwave observations (0.3–500 mm). Because of the translucency of the surface at these wavelengths, the observed microwave intensities depend on subsurface

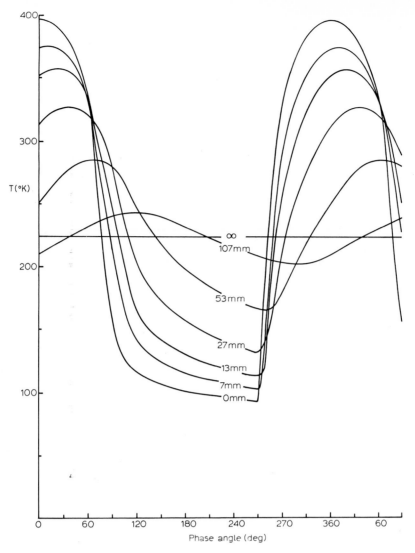

Fig.2. Surface and subsurface temperatures, T, at points in the lunar thermal equator, computed as a function of the phase. The numbers by the individual curves refer, in each case, to the depth below the surface for which the computation was made. A value of $\gamma = 1,000$ has been assumed, and the surface value $z = 0$ in this graph should therefore be identical with the curve labelled 1,000 in Fig.1. (From unpublished calculations by D. O. Gough; see also BASTIN and GOUGH, 1969.)

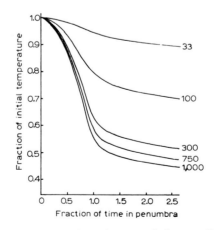

Fig.3. Computed fall in temperature at a point on the lunar thermal equator during an eclipse of the Moon. The temperatures have been normalised to unity at the start of the eclipse. The figures by each curve refer to the assumed value of γ. Apparently, no calculation of the rise of temperature at the end of an eclipse for models having different thermal inertias appears in the literature.

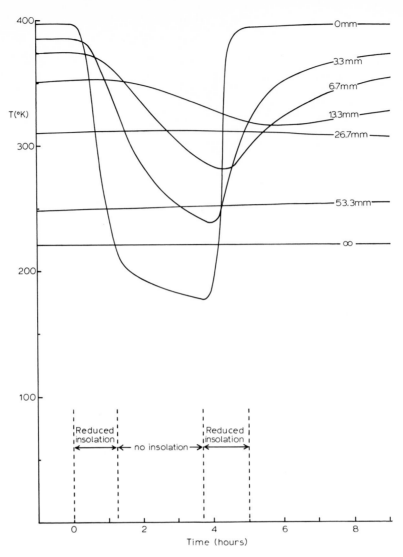

Fig.4. Computed surface and subsurface temperatures, T, at points in the lunar thermal equator, during an eclipse of the Moon. The numbers by the individual curves refer, in each case, to the depth below the surface. A value of $\gamma = 10^3$ cm^2 sec$^{\frac{1}{2}}$ K$^\circ$ cal^{-1} has been assumed in each case. (From unpublished calculations by D. O. Gough; see also BASTIN and GOUGH, 1969.)

TABLE I

MEASURED THERMAL PARAMETERS OF RETURNED APOLLO 11 MATERIAL

Authors	Method	Type	Ambient conditions		k	ρ	c	γ
			gas pressure (torr)	temperature (°K)	(cal. c.g.s.)	(g/cm³)	(cal. c.g.s.)	(c.g.s.)
BIRKEBAK et al. (1970)	line source	D	10^{-3}	205	$4.1 \cdot 10^{-6}$	1.265	0.135	1,200
				299	$5.0 \cdot 10^{-6}$	1.265	0.178	940
				404	$5.8 \cdot 10^{-6}$	1.265	0.202	820
BOWELL (this Chapter)	thermal comparator	A	760	293	$1.6 \cdot 10^{-3}$	3.4	0.176	32
		C	760	293	$0.66 \cdot 10^{-3}$	2.56	0.176	58
HORAI et al. (1970)	modified Ångstrom	A	760	290	$3.87 \cdot 10^{-3}$	2.93	0.174	22
		C	760	290	$1.66 \cdot 10^{-3}$	2.29	0.174	40

temperatures and, thus, are important in the present discussion.

Again, with a plane homogeneous model it can be shown that the intensity I_λ seen from the lunar surface by a distant observer is:

$$I_\lambda = (1 - R_\xi) \int_{z=0}^{\infty} K_\lambda \rho \sec \zeta \, B_\lambda (T_z) \, e^{-k_\lambda \rho \sec \zeta} dz \quad (7)$$

where ζ is the zenith angle of the observed element of the Moon, R_ξ the reflection coefficient at this angle and K_λ the mass electromagnetic absorption coefficient at wavelength λ. If we put the values of T_z given by the conduction theory in eq.7 then comparison of this expression with microwave observations shows that, in the wavelength range above 1 mm:

$$K_\lambda = a\lambda^{-1} \quad (\lambda > 1 \text{ mm}) \quad (8)$$

where a is a constant of value 0.4 $g^{-1}cm^3$.

Below 1 mm wavelength direct measurements on lunar samples (BASTIN et al., 1970) show that the absorption takes the form:

$$K_\lambda = b\lambda^{-2} \quad (1 \, mm > \lambda > 0.1 \, mm) \quad (9)$$

with $b = 2.4 \cdot 10^4 \, g^{-1}cm^4$.

The physical reasons for this variation in electromagnetic absorption are not clear. A single lattice absorption in the far infrared would be expected to give an absorption governed simply by eq.9.

Finally, eq.7 shows that, at long wavelengths, (in practice above about 10 cm) observations give no measurable change in the radiant intensity throughout either an eclipse or a lunation. This is simply because, at these wavelengths, the mean free path of photons is very much greater than the depth of penetration of the thermal wave, so that only an insignificant fraction of the radiation received at the Earth comes from layers in which the temperature varies appreciably with time. At still longer wavelengths ($\sim$ 50 cm) Earth-sited observations allow temperature measurements at considerable depth within the lunar crust. The possibility of discovery of a heat flux resulting from internal heat sources will be discussed later (see p.150).

Temperature-dependent models and radiative conductivity

The thermal conductivity and specific heat of granular substances increase with temperature: in both cases this was predicted theoretically and established experimentally. MUNCEY (1958) pointed out that, as a result, we should expect the mean surface temperature

of the lunar surface throughout a lunar cycle to be less than the temperature of subsurface layers determined from microwave observations; and this is indeed observed. The surface layer can be regarded as a thermal rectifier, transmitting heat into the sub-surface layers during hot daytime conditions, but impeding its outward flow during the night. In Fig.5 the mean subsurface temperature is plotted as a function of wavelength in the microwave region; and the rectifying effect is shown clearly. It follows from eq.7 and 8 that the effective depth of observation increases linearly with wavelength so that, in effect, Fig.5 shows the variation of mean temperature with depth below the surface.

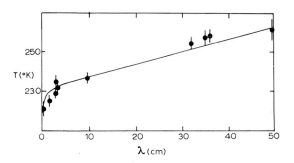

Fig.5. Variation of mean equatorial observed microwave temperature, T, with wavelength λ (After LINSKY, 1966.)

Various authors from WESSELINK (1948) onwards have suggested that a variation of conductivity with temperature could be expected since, in a fragmental material, radiation as well as true lattice condition mechanisms are both effective in transferring heat.

Three possible modes of heat transfer are possible and are shown in Fig.6. Early authors assumed implicitly that the solid was opaque to those infrared wavelengths ($\lambda \sim 10$ cm) at which most of the radiation from a body at the temperature of the lunar crust is emitted. From this assumption they deduced a value of about 30 μ for the mean particle size of the surface layer. However, CLEGG et al. (1966) showed that this result was an oversimplification since photon transport through the individual particles must also be taken into account (see Fig.6). Such a process could occur in crystalline rocks containing no voids, and the estimate of apparent particle size is, rather, an estimate of the photon mean free path. Recent measurements on lunar type D material (BASTIN et al., 1970) gave a value of 300 cm^{-1} for the absorption coefficient at 10 μ—a result in excellent agreement with Wesselink's estimate if the alternative purely radiative process is considered effective. For such a medium the total effective conductivity K can be written:

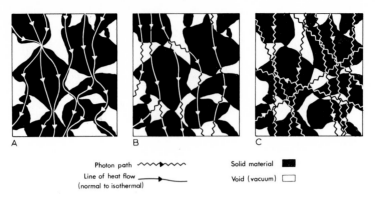

Fig.6. Possible modes of heat transfer in porous lunar fines. A. Pure conduction. B. Conduction–radiation. C. Pure radiation.

$$K = K_l + \frac{16\sigma}{3} \cdot \frac{T^3}{\overline{K}(T)} \qquad (10)$$

In this expression, deduced by CLEGG et al. (1966), the first term K_l represents the lattice contribution. In a particulate medium this term will be greatly reduced by the effect of the very small particle contacts. Thus although the conductivity of typical crystalline rocks is about $3 \cdot 10^{-3}$ cal./cm · sec · K °, particulate media in vacuo have a pure conduction contribution of about 10^{-6} cal./cm · sec · K °. It is for this reason that the radiative (final) term in eq.10 is important: this term contains $\overline{K}(T)$, a mean (Rosseland) electro-

magnetic absorption coefficient, weighted over the spectral range according to the black body distribution of radiation at the temperature T. Earth-based observations of lunar thermal radiation show the two terms to be almost equal at $350°K$. As a result we can estimate $\overline{K}(T)$; this estimate is in close agreement with the Earth-based measurements of lunar samples.

The variation of conductivity with temperature is well illustrated by eclipse measurements in the 8–14 μ region which correspond to observations of the surface temperature. These are shown in Fig.7, together with some predictions from a "particulate thermophysical" model considered by WINTER and SAARI (1969). The initial, slow decrease in the surface temperature of the Moon at the beginning of an eclipse can be accounted for well by introducing a conductivity which increases with temperature, whilst the fact that the temperature dwells at an almost constant value throughout the centre of the eclipse strongly indicates a conductivity which increases rapidly with depth.

Effects of roughness on thermal properties

A small, plane element of a rough surface will radiate as much heat as a similar element of a smooth surface with the same temperature and albedo. However, in general, the element of the rough surface will also receive radiation from other surface elements, and the angle the element makes with the Sun will not be a simple function of the lunar phase or location of the site considered. In addition, there are complex shadowing problems. With a given model for the surface configuration the calculation of the temperatures at all points and for all times throughout eclipses and lunations is a complex problem in which the use of high speed computers is almost essential. The equations to be solved are essentially the analogues of eq.1 and 2. Eq.1 must be replaced by its three-dimensional equivalent:

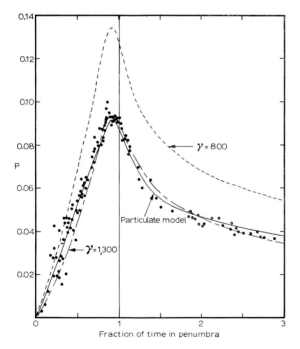

Fig.7. Thermal measurements (filled circles) at 8–14 μ during an eclipse together with the predictions (continuous curve), for the centre of the lunar disk, of the "Particulate Thermophysical Model" of WINTER and SAARI (1969). The quantity P is a complicated "differential energy parameter". It is clear that the particulate model gives better agreement with observation than either of the smooth homogeneous models (broken curves).

$$\frac{\partial T}{\partial t} = \frac{k}{\varrho c} \nabla^2 T \qquad (11)$$

whilst eq.2 must be modified to include radiation from other surfaces:

$$(1 - \alpha_1)(I + R) = (1 - \alpha_2)\sigma T^4 - k\frac{\partial T}{\partial n} \qquad (12)$$

Here, n is measured in the direction normal to the surface and R, the radiation reaching the element from all other surfaces, is a function of the solid angle subtended by every surface element radiating to the element in question and depends also on the temperature of the radiating elements.

The various models which have been used to investigate the thermal properties of a rough surface are shown in profile in Fig.8. The reasons for adopting

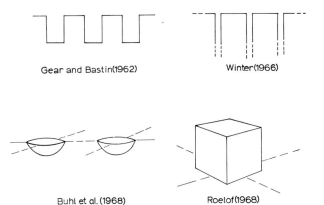

Gear and Bastin(1962)   Winter(1966)

Buhl et al. (1968)   Roelof(1968)

Fig.8. Rough surface models whose thermal properties have been computed.

these particular models include both the desire for computational simplicity and an attempt to reproduce, in the model, known features of the terrain. It is found that two length scales characterise the problem. These are:

$$L_1 = \left(\frac{2k}{\rho c \omega}\right)^{\frac{1}{2}} \quad \text{and} \quad L_2 = \frac{k}{\sigma T^3} \qquad (13)$$

These two lengths become approximately equal during the lunar night but, during the day, L_1 is of the order of several metres whereas L_2 is about 1 mm. If the roughness scale size were very much less than L_2 then the whole surface would be isothermal and the roughness would not affect the thermal situation (except in as much as it might decrease the surface albedo). For roughness scales greater than L_1 during the daytime eq.12 again simplifies to a one-dimensional equation. With roughness scale sizes between L_1 and L_2 the situation is complex and can be solved only by computational means. This is, however, an interesting

region since radar measurements and the more recent stereoscopic Apollo photography have shown a considerable degree of roughness at these dimensions.

A large number of the global lunar features can be explained only by roughness on a centimetre or millimetre scale. These include the result that the subsolar point has an apparent temperature (measured from the radiation it emits) which varies with the angle of observation. As another example, the poleward and equatorial scans of the full Moon show less limb darkening than would be expected from the plane model. These cases are particular examples of the radiant intensity from a point on the Moon: on any model these must be considered a function of phase, latitude, longitude and angle of observation and, thus, it is difficult to show a comparison between observation and prediction on a single diagram. However, the observation and the predictions of various models are shown in Fig.9 and 10 for the two particular cases cited above. As a further example, the poleward variation of subsurface temperatures which are determined from microwave measurements are again exactly accounted for only by a rough model: the smooth model gives a poleward darkening function which is considerably less rapid than that observed. Rough models may

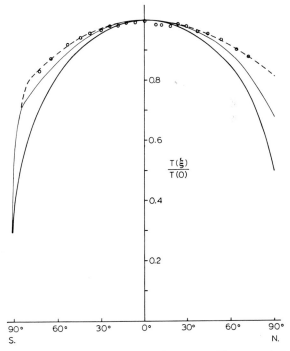

Fig.9. Equatorial temperatures at full Moon. The open circles refer to the measurements of SAARI and SHORTHILL (1967). The thick, continuous curve indicates the computation for the plane model. The thin, continuous line and broken curve are for two different trough models (see Fig.8), both with horizontal and vertical surfaces. The broken curve is in good agreement with observation.

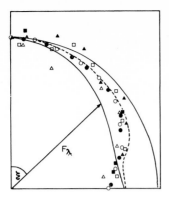

Fig.10. Normalised radiant intensity, F_λ, at wavelength 11 μ at the subsolar point as a function of the zenith angle, ξ, of observation. Open and filled symbols refer, respectively, to values obtained from the waxing and waning Moon: triangles refer to Pettit and Nicholson (1930); squares to Geoffrion et al. (1960); and circles to Saari and Shorthill (1967). For further details see Bastin and Gough (1969).

also account for the high, daytime microwave temperatures which have been observed and may also be responsible for some, but not all, of the "hot spot" observations of Saari and Shorthill (1967).

The roughness having the greatest physical effect in all these cases is in the centimetric and millimetric scale region. On a larger scale roughness could, in theory, produce important effects; but appreciable roughness does not exist on such a scale. On a sub-millimetric scale the isothermal result discussed above means that roughness is unimportant. On a millimetric and centimetric scale a mean slope of about 30° to the horizontal is implied by thermal measurements, although models with vertical and horizontal sides give the best agreement with observation.

Localised thermal anomalies

Fig.11 shows a scan across the Moon made by Shorthill and Saari during an eclipse using a detector responding to radiation in the 8–14 μ region. Using the value $(k\rho c)^{-\frac{1}{2}} = 10^3$ cm$^2 \cdot$ sec$^{\frac{1}{2}}$(K°)$^{+1} \cdot$ cal.$^{-1}$ one

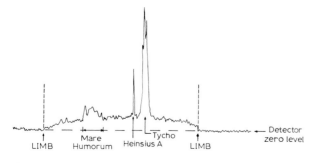

Fig.11. Scan across the Moon at 8–14 μ wavelengths made during totality of the eclipse of December 19, 1964.

would expect the surface temperature to drop to about 170°K by the end of totality of the eclipse. Whilst this was found over most of the scan, a number of other regions—"hot spots"—had temperatures of up to 60°K above the predicted figure. The 8–14 μ region is an extremely sensitive region for such observations since, in this wavelength range, radiation from a surface varies very rapidly with its temperature (Wien regime). For example, the hot spots are only about 35% warmer than their surroundings, yet the radiation detected in the 8–14 μ atmospheric window can be greater by a factor of three or more (see Fig.11). Hot spots are associated with craters but also occur in maria. During the lunar day the more intense spots are cooler than their surroundings by a few degrees Kelvin. For this reason internal heat sources cannot be responsible for most of the anomalies and it now seems almost certain that most of them are caused by large rocks, or exposures of rock, on the surface. Such rocks will clearly have considerably higher thermal conductivity than that of the mean regolith and, with an effective thermal inertia of about 30 c.g.s. units, will cool much less during eclipses (see Fig.4). An alternative hypothesis that the hot spots are due to preferential roughness can be invoked to account for the required enhancements during an eclipse but also requires a higher daytime temperature. It therefore seems that such a mechanism cannot account for the most intense hot spots. However, Hunt et al. (1968) have discovered hot spots with daytime enhancements and the roughness hypothesis may account for many of the less intense spots (excess temperature during eclipse ≤ 15°K). A histogram of the number of spots as a function of their temperature excess shows a discontinuity at about 15°K suggestive of two different effective mechanisms.

Similar anomalous regions have been found from temperature measurements during the lunar night. The work of Allen and Ney (1969) is particularly revealing in this respect since it has been made at a number of wavelengths in the range 3–15 μ and, in this way, it has been shown that the surfaces of three rayed craters—Tycho, Copernicus and Aristarchus—have a composite thermal structure. In Tycho, for example, the spectrum during the lunar night corresponds to that predicted from a surface one tenth of which is at a temperature of 200°K and the rest at 100°K. The figure of 200°K suggests solid rock although the virtual constancy of the temperature cannot be accounted for by small rocks, since they have too low a thermal capacity in comparison with their surface area. Large rock outcrops are suggested and this result is supported by an analysis of Lunar Orbiter photography by G. Fielder (personal communication, 1970) and also by a detailed study of the daytime radiation from this crater (BASTIN and GOUGH, 1969). In conclusion, it now seems clear that spatial change in thermal inertia of the lunar surface from place to place is the most important cause of thermal anomalies. The presence of composite surfaces composed of areas of widely differing thermal inertia is strongly indicated, whilst the effects of differential roughness and of spatial changes in the albedo for incoming and outgoing radiation have considerable—although not the most dramatic—effects.

The regolith and the deeper layers

Measurements on lunar samples returned to Earth have given mean values for the fines (type D) conductivity of $k = 2 \cdot 10^{-6}$ cal./cm $\cdot$ sec $\cdot$ (K°), density $\rho = 1.65 \pm 0.5$ g/cm^3, specific heat $c = 0.06$–0.20 cal./g $\cdot$ K°. These give a value of the thermal inertia of about 10^3 cm$^2 \cdot$ sec$^{\frac{1}{2}} \cdot$ (K°) $\cdot$ cal.$^{-1}$. This is in agreement with the previous, Earth-based, infrared results. The solid, crystalline rock has values $k = 3 \cdot 10^{-3}$ cal./cm $\cdot$ sec $\cdot$ (K°), $\rho = 3$–4 g/cm^3, $c = 0.06$–0.20 cal./g $\cdot$ K° giving $(k\rho c)^{-\frac{1}{2}}$ around 25 cm$^2 \cdot$ sec$^{\frac{1}{2}} \cdot$ (K°) $\cdot$ cal.$^{-1}$. In addition to the variation of conductivity with temperature a very marked increase of specific heat with temperature has also been observed. The breccia samples have somewhat lower values of the thermal conductivity.

Estimates (see, for example, Chapter 1) show that the bedrock starts at distances of from ~ 1 to ~ 10 m below the surface of the regolith. This bedrock presumably has thermal parameters equal to those of the crystalline rocks collected at the surface since these rocks probably originate from the bedrock. What is not clear is how the conductivity varies in the lower regolith layers below the top surface. With increasing depth there is an increasing degree of compaction and, also, the possibility of an appreciable gas pressure in the lower layers. Both of these factors will give a thermal conductivity which increases with depth. However, the exact form of the variation is of considerable importance since it determines the mean temperature difference to be expected in the top layers as a result of any internal heat sources. For a given heat flux, and in the steady state, the temperature difference across any layer will be inversely proportional to the thermal conductivity of the layer. For a central lunar temperature of about $1,000$°K this would imply a temperature gradient of 1° per metre if the layers had a conductivity of 10^{-6} but very much less if the conductivity

approached that of the crystalline bedrock. This temperature gradient could be measured by microwave observations at different wavelengths. There have been several Russian measurements which support a considerable temperature gradient (TROITSKI, 1967; see also Fig.5). A somewhat lower value is implied by BALDWIN's (1961) measurements.

Thermal conductivity measurements of lunar materials returned to Earth[1]

Methods of determining thermal conductivity

In the last few decades, many methods of determining the thermal conductivity of materials in solid and powder form have been developed—mainly for industrial, nuclear and geophysical purposes. Few of these methods are applicable to a study of returned lunar samples. There are three principal experimental difficulties in the study of lunar materials:

(1) They are, perforce, in short supply. They should not be contaminated, destroyed or, in so far as is possible, cut and re-shaped.

(2) They are likely to be texturally inhomogeneous, possessing sharp local variations in conductivity.

(3) It would be ideal to make measurements in an environment approximating to lunar surface conditions: that is, with an ambient gas pressure of the order of 10^{-13} torr and over the temperature range 100–400°K. In the case of the fines some attempt should be made to reproduce the lunar surface bulk density of about 1 g/cm^3, and to study the effects of compaction.

Generally speaking, steady state methods, such as radial or longitudinal heat flow measurement, yield accurate and reproducible thermal conductivity values but require large, carefully surfaced samples and are slow. Commonly used transient methods require smaller samples and have a variable heat input. They are fast and reproducible but not as accurate as steady state methods. Usually it is the thermal diffusivity (κ) and not the conductivity ($k = \kappa\rho c$) which is derived, so that the specific heat and density of a test material have to be found independently. In the lunar materials available to the author these quantities vary locally and it is not practicable to study them.

It is necessary, then, to adopt a non-steady state method which will deal with small samples and which, in preference, is quick in operation and yields the

---

[1] This section is written by E. L. G. Bowell.

conductivity directly. Techniques which at least partially fulfil these requirements have been used by three of the Apollo 11 principal investigators, and are described below.

The line source heating method

The line source heating method, such as described by WOODSIDE and MESSMER (1961) consists of a thermal probe—a heating wire of very low temperature coefficient of resistance—and a thermojunction embedded alongside it in the sample under test. A constant voltage applied across the heating wire raises the sample temperature locally by an amount which can be related directly to the sample conductivity. This technique was modified by BIRKEBAK et al. (1970) to measure the conductivity of lunar fines contained in a Teflon cell measuring internally $2.5 \times 1.3 \times 1.3$ cm³. A Nichrome heating wire, 0.0203 cm in diameter, was used both as the line source and as resistance thermometer to measure the temperature of the rock powder. An iron-constantan thermocouple monitored the temperature in the powder at a distance of 0.02 cm. The Teflon cell was placed in a vacuum chamber and surrounded by a blackened copper shroud which could be heated or cooled to control the ambient temperature over the range 200–400°K. Ambient gas pressure was generally less than 10^{-3} torr.

The modified Ångstrom method

The modified Ångstrom method described by KANOMORI et al. (1968, 1969) measures the thermal diffusivity of a small cylinder or rectangular prism of test material. Four Apollo 11 samples—two crystalline (type A) and two breccias (type C)—have been investigated by HORAI et al. (1970). Each sample is in the form of a rectangular prism measuring $1 \times 1 \times 2$ cm³, the end face of which is pasted on to a nickel base wound with a heating wire. A 5- to 20-sec period square wave electric current heats the nickel base, the geometry of which ensures that the higher harmonics are attenuated so that the sample interface is subjected to an almost sinusoidal temperature wave. Thermocouples monitor the temperature variations at both ends of the sample. A determination of the ratio of the temperature wave amplitudes and the phase shift across the sample yields the thermal diffusivity.

This method is particularly suitable for making diffusivity estimates over a wide range of temperatures; the whole apparatus is small and robust and can be operated easily in a furnace. Disadvantages are that there are unknown edge effects, due to the presence of a substantial radiative component of heat transport at high temperatures and sensitivity to thermal resistance at the thermojunction sites. Horai et al. measured the diffusivity of rocks in the temperature range 150–420°K and found a linear relation between the reciprocal of the thermal diffusivity and temperature. The respective diffusivity values for the two crystalline samples, and the two breccias, were substantially the same; but the ratio of diffusivities crystalline/breccia was found to be 1.7 at all temperatures.

The thermal comparator method

The so-called thermal comparator method, first proposed and developed by POWELL (1962) at the National Physical Laboratory was used here to measure the thermal conductivities of crystalline (type A) and breccia (type C) lunar materials: preliminary results were reported by BASTIN et al. (1970). The principle of operation of the comparator is based on the common observation that, under ambient conditions, good conductors feel cooler to the touch than poor ones. The method has the great advantage that, once the comparator is calibrated, thermal conductivity may be found directly. Another advantage, where invaluable lunar rocks are concerned, is that poor conductors having volumes as small as 0.5 cm³ may be used.

The apparatus is shown in Fig.12. The comparator consists of two top-shaped copper blocks, each weighing about 10 g, housed in a perspex block so that only their rounded tips protrude (see Fig.13). The two comparator blocks are joined with thin copper wire; constantan leads run from thermojunctions at their tips so that the e.m.f. developed across the constantan leads is a measure of the temperature difference between the tips of the comparator blocks.

When in its housing, the comparator is raised into a chamber (Fig.12) and heated electrically to a uniform temperature of about 40°C above the sample temperature of 20°C. Then it is lowered so that one of the comparator blocks makes good thermal contact with the sample. A perspex peg on the comparator housing prevents the other block from touching the sample. Heat flows by conduction from one comparator block into the sample; the temperature difference set up in the comparator is monitored for about 30 sec to determine the rate of flow of heat. Typical cooling curves, for various materials, are shown in Fig.14. It can be seen that, for the breccia, a temperature difference of about 90 μV (about 2.2°C) is registered in the comparator after 12 sec. After repeated trials this may be determined to within 1 μV. Now the amount of heat lost by the comparator to the sample depends on several factors. These include: (1) the thermal con-

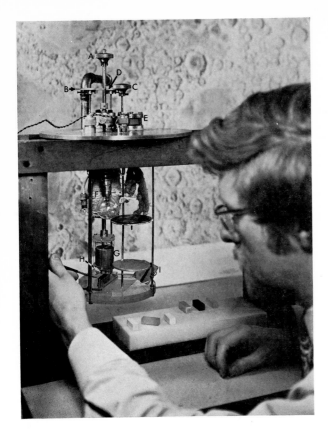

Fig.12. Thermal comparator apparatus used at the University of London Observatory to determine the thermal conductivity of returned lunar samples. The apparatus is placed in a 5-l brass chamber to work at pressures of less than 10^{-3} torr. A = Control for vertical movement of comparator; B = control for adjustment of comparator perpendicular to axis of sample (author is pointing at lunar sample); C = control for heating chamber lid and radiation guard; D = control for adjustment of comparator along axis of sample; E = viewing port; F = lagged heating chamber; G = comparator load; H = comparator in perspex housing; I = mirror for viewing sample when equipment is operating in vacuo. (Photo: courtesy of Turner Asbestos Cement Company Ltd.)

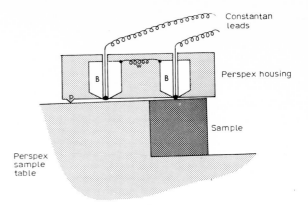

Fig.13. The two copper comparator blocks, B, in their perspex housing, and the sample table. The symbol w refers to a thin, copper wire; p is a peg for support of the comparator.

CLARK and POWELL (1962) and KOLLIE et al. (1964) have found that, for materials with conductivities below about $k = 0.02$ cal./cm · sec · (K°)—consolidated lunar materials have conductivities an order of magnitude less—the temperature difference set up in a comparator after a short time varies almost linearly with the square root of the conductivity.

The comparator was calibrated using materials with thermal conductivities known to straddle the conductivities expected of the returned lunar samples. These "standards" were carefully shaped and finished to resemble the $3 \times 1 \times$ cm^3 saw-cut lunar materials. By measuring the differential temperature (ΔV) set up in the comparator, with the comparator at a temperature of 40 °C above the sample temperature and a contact period of 12 sec, it was possible to plot an

ductivity of the sample; (2) the initial temperature difference between comparator and sample surface; (3) the area of contact at the comparator tip; (4) the load imposed at the comparator tip; (5) the size and surface texture of the sample; and (6) the atmosphere in the chamber.

Provided that the experimental conditions are maintained so that factors (2) to (6) do not change, the temperature-time records (as in Fig.14) may be used directly to infer the sample's thermal conductivity given that the instrument has been calibrated. This is so because the method is a rapid and transient one in which temperature gradients are essentially localised. Other potentially troublesome effects, such as variation in sample emissivity and convection of gas in the chamber, may be neglected in the first instance.

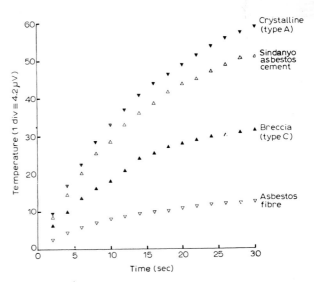

Fig.14. Typical comparator cooling curves for various materials including lunar crystalline and breccia samples. An initial difference of 40 °C between the comparator and each sample was used, with a chamber containing air at 1 atm and 290 °K.

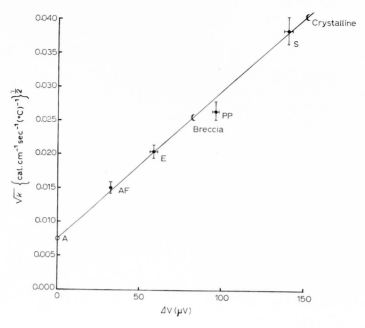

Fig.15. A thermal comparator calibration curve. The temperature differential, ΔV, in the comparator—using an initial temperature difference of 40°C between the comparator and samples and a contact period of 12 sec—is plotted against the square root of the thermal conductivity, k. A = air; AF = asbestos fibre; E = ebonite; PP = plaster of Paris; S = Sindanyo asbestos cement.

instrumental calibration curve giving $\sqrt{k}$ as a function of ΔV. Hence an unknown conductivity could be derived directly from this curve using only a measurement of ΔV.

Fig.15 illustrates the results for some of the standard materials used to calibrate the comparator; included on the calibration line are the measurements made on the lunar samples. Error bars in $\sqrt{k}$ for the standard materials are estimated errors and those in ΔV are standard errors of the mean of six or more measures. It will be noted that the calibration line very nearly intercepts the conductivity axis at the value of $\sqrt{k} = 7.5 \cdot 10^{-3}$ (cal. c.g.s.)$^{\frac{1}{2}}$ for air at 290°K and 1 atm pressure.

Results of thermal conductivity measurements on lunar samples

The results of the three experiments described above are set out in Table I. The values for specific heat are taken from the work of ROBIE et al. (1970), who investigated vesicular basalt (type A) and fines (type D) samples. They found substantially similar specific heats for the two materials, ranging from 0.06 cal./g · °C at 100°K, to 0.20 cal./g · °C at 400°K. The mean conductivities and thermal inertias are given for each pair of Horai's samples.

References

ADAMS, J. B. and JONES, R. L., 1970. *Science*, 167: 737.
ALLEN, C. W., 1962. *Astrophysical Quantities*, 2nd edition. Athlone Press, London.
ALLEN, D A. and NEY, E. P., 1969. *Science*, 164: 419.
BALDWIN, J. E., 1961. *Monthly Notices Roy. Astron. Soc.*, 122: 513.
BASTIN, J. A. and GOUGH, D. O., 1969. *Icarus*, 11: 289.
BASTIN, J. A., CLEGG, P. E. and FIELDER, G., 1970. *Science*, 167: 728.
BIRKEBAK, R. C., CREMMERS, C. J. and DAWSON, J. P., 1970. *Science*, 167: 724.
BUHL, D., WELCH, W. J. and REA, D. G., 1968. *J. Geophys. Res.*, 73: 5281.
CLARK, W. T. and POWELL, R. W., 1962. *J. Sci. Instr.*, 39: 545.

CLEGG, P. E., BASTIN, J. A. and GEAR, A. E., 1966. *Monthly Notices Roy. Astron. Soc.*, 133: 63.
FRYXELL, R., ANDERSON, D., CARRIER, D., GREENWOOD, W. and HEIKEN, G., 1970. *Science*, 167: 734.
GEAR, A. E. and BASTIN, J. A., 1962. *Nature*, 196: 1305.
GEOFFRION, A. R., KORNER, M. and SINTON, W. M., 1960. *Lowell Obs. Bull.*, 5: 1.
HARRIS, D. L., 1961. In: G. P. KUIPER and B. M. MIDDLEHURST (Editors), *Planets and Satellites*. University of Chicago Press, Chicago, Ill., p.272.
HORAI, K. I., SIMMONS, G., KANAMORI, H. and WONES, D., 1970. *Science*, 167: 730.
HUNT, G. R., SALISBURY, J. W. and VINCENT, R. K., 1968. *Science*, 162: 252.

JAEGER, J. C., 1953a. *Aust. J. Phys.*, 6: 10.

JAEGER, J. C., 1953b. *Proc. Cambridge Phil. Soc.*, 49: 355.

KANAMORI, H., FUJII, N. and MIZUTANI, H., 1968. *J. Geophys. Res.*, 73: 595.

KANAMORI, H., MIZUTANI, H. and FUJII, N., 1969. *J. Phys. Sci.*, 165: 1211.

KOLLIE, T. G., MCELROY, D. L., GRAVES, R. S. and FULKERSON, W., 1964. *Nuclear Metallurgy*, 10: 651.

KROTIKOV, V. D. and SHCHUKO, O. B., 1963. *Soviet Astron.*, 7: 228.

LINSKY, J. L., 1966. *Icarus*, 5: 606.

MUNCEY, R. W., 1958. *Nature*, 181: 1458.

MURRAY, F. H., 1965. *J. Geophys. Res.*, 70: 4959.

PETTIT, E. and NICHOLSON, S. B., 1930. *Astrophys. J.*, 71: 102.

POWELL, R. W., 1962. *Final Report, Office of Naval Research*, Contract No. (G)-0064-62.

ROBIE, R. A., HEMINGWAY, B. S. and WILSON, W. H., 1970. *Science*, 167: 749.

ROELOF, E. C., 1968. *Icarus*, 8: 138.

SAARI, J. M. and SHORTHILL, R. W., 1967. In: *Physics of the Moon—Sci. Technol. Ser., Am. Astronaut. Soc.*, 13: 59.

SMART, W. M., 1960. *Spherical Astronomy*. Cambridge University Press, London.

TROITSKI, V. S., 1967. *Proc. Roy. Soc. London, Ser. A*, 296: 366.

WESSELINK, A. J., 1948. *Bull. Astron. Inst. Neth.*, 10: 351.

WINTER, D. F., 1966. *Intern. J. Heat Mass Transfer*, 9: 527.

WINTER, D. F. and SAARI, J. M., 1969 *Astrophys. J.*, 156: 1135.

WOODSIDE, W. and MESSMER, J. H., 1961. *J. Appl. Phys.*, 32: 1688.

Subject Index

Absorption within the regolith, 128, 140
Alaska, 99
Albedo, 43, 45, 49, 55, 73, 75, 80, 88, 100, 115, 125–132, 140, 143, 149, 151
Alpha scattering device, use of the, 73, 135
Alphonsus, 4, 5, 9, 45
Aluminium, 8, 73, 141
Andesite, 73
Angle of emergence, definition of, 115
– of incidence, definition of, 115
– of vision, definition of, 115
– – –, relation with the degree of polarisation, 125, 126, 129, 131
– – –, relation with brightness surge, 121, 122
Ångström method, modified, 152
Anorthosite, 12, 73, 135
Apollo VIII, 2, 11
– X, 2, 11
– XI, 2, 3, 10, 12, 73, 122, 123, 125, 131, 135–137, 140, 146, 152
– XII, 2, 7, 10, 12, 42, 73, 122, 137
– XIII, 12
– XIV, 2
– XV, 2
Archemedian Series, 38, 93
Arcuate ridging of flows, 70
Aristarchus, 27, 46, 55, 76–93, 100, 112, 126, 141, 150
–, debris channels of, 81
–, drainage channels of, 78
– Plateau, 27, 37, 38
–, ray system of, 92
–, transient phenomena around, 27, 91
Arizona, 129
Azimuthal angle, definition of, 115

Barium, 135, 137
Basalt, 125, 130, 131, 135, 137, 138, 140
Basaltic achondrites, 135
Base surge hypothesis, 89
– – deposits, 90, 92, 98
Benches at the crater floor, 85, 86, 88, 91, 100
Breccia, examination of, 10, 135, 136
–, comparison between lunar and terrestrial, 100
Brightness, 115–122
–, errors in, 116, 117
– measurements, 116, 118, 120
– surge, relation with the angle of vision, 121, 122
Bronzite, 130

Cadmium, 135
Caesium, 135
Calcium, 135
Calderas, 27, 38, 41, 43, 101, 105, 113
–, collapse of, 93
Canada, 100
Carbide, 136
Carbon, 135–137, 140
Catalogue of Fedoretz, 115, 121
Cauchy domes, 43
Chemical composition of lunar rocks, 73, 135, 136
Chile, 129
Chlorine, 141
Chondrites, 10

Clearwater Lake crater, 100
Coalescing flows, 88
Cobalt, 137
Collapse depressions, 36, 62, 66, 78, 80–82, 88, 111, 112
Compaction of surface layers, 100, 123, 131, 151
Convection currents, 46
Copernican Era, 38, 49
Copernicus, 1, 93, 150
Copper, 135
Cosmic ray nuclides, 12
Crater alignments, 108, 110
– chains, 1, 3, 11, 17, 98, 101
– counts, 18, 73–75, 85, 87, 89, 90, 91, 106, 107, 108
– diameter, 45, 67, 74, 87
– floors, isostatic uplift of, 91
Craters, breached, 67, 70, 71, 89
–, circularity of, 113
–, collapse, 25, 34, 46, 50
–, dark halo, 4, 5, 10, 45
–, denudation chronology of, 93, 94
–, dimple, 3, 4, 25, 81
–, double, 23
–, endogenic, 1, 25, 49, 90, 108, 111, 112
–, eumorphic, 1, 22–24, 72, 74, 85, 87, 90
–, explosion, 25, 45, 75, 113
–, ghost, 44, 45
–, host, 106, 107
–, internal, 111, 112
–, meteoritic, 93
–, mounds in, 23
–, number density of, 19–21, 27, 49, 62, 67, 85, 87, 90, 91, 98, 101, 102, 106, 108, 111, 112
–, obliteration of, 106–108, 112
–, parasite, 106, 107
–, Poisson distribution in relation to, 106, 108, 110
–, primary impact, 22, 23, 25, 41, 49, 75, 87, 88, 90, 95, 107
–, ray, 21, 100
–, rille, 45
–, secondary impact, 1, 3, 18–23, 38, 41, 49, 73–75, 87, 89, 92, 93, 98, 102, 112
–, shape of, 21, 22
–, submorphic, 1, 25, 106
–, volcanic, 25, 32, 45
Creep, 95, 102
Crushing strength of rocks, 8
Curium, 135

Darkening of the surface layer by proton irradiation, 131, 132, 137–140
Density of lunar rocks, 8, 12, 22, 73, 151
Denudation processes, 43, 128, 130, 132
Distribution of craters, 106–114
Domes, 27, 38, 43, 47–50, 70, 73

Eclipse, lunar, 144–150
Eclogite, 22
Ejecta blanket, 93, 98–101
– –, grooves in the, 99
– –, hummocks in the, 98
– flow, 98–100
Electromagnetic absorption coefficient, 147, 148